Springer Praxis Books

More information about this series at http://www.springer.com/series/4097

Colin Tucker

How to Drive
a Nuclear Reactor

 Springer

Colin Tucker
Wickham Market, Suffolk, UK

SPRINGER PRAXIS BOOKS IN POPULAR SCIENCE

Springer Praxis Books
ISSN 2626-6113   ISSN 2626-6121   (electronic)
Popular Science
ISBN 978-3-030-33875-6   ISBN 978-3-030-33876-3   (eBook)
https://doi.org/10.1007/978-3-030-33876-3

This Springer imprint is published by the registered company Springer Nature Switzerland AG.
The registered company address is: Gewerbestrasse 11, 6330 Cham, Switzerland

To Lynette

Preface and Acknowledgements

Have you ever wondered how a nuclear power station works? This book will show you, by asking you to imagine that you're a trainee reactor operator on a Pressurised Water Reactor (PWR), the most common type of nuclear reactor in the world. It'll take you on a journey from the science behind nuclear reactors, through their start-up, operation and shutdown. Along the way, it covers a bit of the engineering, reactor history, different kinds of reactors and what can go wrong with them. This book will show you how reactors are kept safe, and what it feels like to drive one.

So what inspired me to write this book? It was a conversation about a book entitled *How to Drive a Steam Locomotive* (by Brian Hollingsworth). I was describing to a friend how the author puts the reader on the footplate of a locomotive and then gradually introduces them to the controls in front of them; what they each do; and what might go wrong. By the end of the book, it felt like you were really there. The conversation ended with me complaining about the lack of any similar book describing nuclear reactors. I have searched for such a book, but have found that most concentrate on energy policy or on nuclear accidents, with only a few short chapters on reactor operation. My experience is that often people want to know more.

So I decided to write *this* book. I hope you enjoy reading it as much as I enjoyed writing it. (I'll let you be the judge of whether or not it matches up to the original.)

As with many industries, nuclear power stations use a lot of jargon. Hopefully, you won't find this too off-putting—there is an Index at the back, which may help. Different kinds of reactors use different jargon (of course!) and you'll see that this book is heavily PWR-biased, though other reactors do make an appearance. Confusingly—especially for people new to the industry—it's not uncommon for power station equipment to have two or more different names, often used interchangeably, especially if that equipment can have different functions at different times. Examples include using the word 'Containment' instead of 'Reactor Building', 'Reactor Coolant System' for 'Primary Circuit', 'Fuel Rod' for 'Fuel Pin', etc. I've tried very hard to only use single terms in this book. To my ear, and perhaps to others who work at PWRs, that makes some of the text feel a little clumsy. Hopefully, to everyone else it will make things clearer. My advice to anyone reading this book is not to get too hung up on the jargon; it's the safe operation of the reactor that matters, not the labelling.

I want to start my acknowledgements by thanking my wife, Lynette, for encouraging me and helping me find the space and time to devote to writing this book. It's not easy to fit this sort of thing into your spare time without other things being displaced. I also need to thank my first readers, Nicholas Butt and Kevin Martin, who provided both technical and non-technical review comments which have (mostly) been addressed. It can't have been easy to read drafts of chapters when you don't have a clear idea of how it's all supposed to fit together. Their patience and perseverance were much appreciated.

I owe an enormous debt of gratitude to the staff of the UK's Sizewell 'B' nuclear power station. This has been my base for nearly 25 years, primarily working in the field of nuclear safety. Most of my experience of PWRs is Sizewell-based, and I accept that there are risks in this for an author; not every PWR is the same. I hope I've been flexible enough in what I've written for those at other PWRs (and indeed, at other reactor designs) not to feel excluded. Sizewell has a marvellously 'open' culture where I've found that I can ask questions on anything to fill gaps in my knowledge. Beyond this, I'd especially mention the support from the Management Team and from EDF Energy Corporate staff with this project. Their enthusiasm for it from the outset, without interfering in any way with its content, has made it so much more achievable.

Finally, I should mention the 'Nuclear Safety Group' at Sizewell B. Their depth of knowledge, experience, willingness to challenge, patience and rigour go a long way to keeping Sizewell 'B' as safe as it is. Their humour makes it enjoyable! This book, though ostensibly concerned with reactor operation, probably comes closest to a view of the world as it's seen from the Nuclear Safety Group. Make of that what you will…

The majority of the content of this book is my own. Where opinions are expressed—and there are a few—they are also mine, and do not in any way reflect the views or policies of EDF Energy or of any other company. That, of course, means that any errors that you find must also be mine. For these, I apologise and say 'well done!' if you've spotted one.

Personally, I find nuclear reactors fascinating. I hope you will too.

Suffolk, UK   Colin Tucker
September 2019

Contents

1

One Man and His Dog

•

I've heard it said that a modern nuclear power station could be operated by one man and a dog. The man would be there to feed the dog, and the dog would be there to bite the man if he touched any of the controls...

If only.

1.1 Reading This Book Won't Qualify You to Drive a Nuclear Reactor

The idea of this book is to explain how a nuclear reactor works and how it can be operated to produce power for the electricity grid. This book won't qualify you to drive a nuclear reactor. That takes a couple of years of training, including hundreds of hours in a simulator. On the other hand, this book will probably give you a much better idea of what is involved in driving one.

So, in starting this book, let's imagine that you've passed all the entrance tests for a job in the 'main control room' of a modern nuclear power station, like the one in Fig. 1.1, and you're ready to learn how it all works. Now your supervisor suggests you change reactor power. Do you have any idea what to do?

Or perhaps the computer system displays an alarm. What does it mean? Which of the quarter of a million or so different items of equipment—depending on how you count them—does this alarm refer to? Is it a problem? Will you respond using one of the few hundred controls and indications in the control room or will someone have to be sent to look at the equipment locally? Could something more significant be going on? Will you have to get

© Springer Nature Switzerland AG 2019
C. Tucker, *How to Drive a Nuclear Reactor*, Springer Praxis Books,
https://doi.org/10.1007/978-3-030-33876-3_1

Fig. 1.1 Part of a PWR control room

ready to shut down the plant. There'll be tens of thousands of possible alarms on a modern station, and just as many procedures to follow.

As a trained reactor operator, you'll need to be able to decide when to act quickly and when to act in a more measured way. Safety is your overriding priority as it is for anyone who drives a reactor or works at a nuclear power station. After safety, you can think about what's best for the people and the plant, but safety comes first. Despite this, you'll understand that your power station is just a factory for making electricity. Unnecessary shutdowns are going to be expensive.

I've never seen an individual book (including this one) that covers everything you'll need to know. After all, an operator has very many controls at their fingertips. This book won't take you through the function of each one, but it will show you how a competent operator works in partnership with both the physics of the reactor and with the automatic systems out on the plant.

So what would make you a successful (and safe) person to drive a nuclear reactor? Well, you're probably going to need some kind of science or engineering background; but this needn't mean a university degree. It could be from an apprenticeship, for example. You'll need the ability to learn a lot about a lot of different things, without necessarily becoming an expert in any of them. You'll need to be rigorous in following procedures, but not blindly—if something doesn't feel right, you should be the first person to stop and ask the question 'Is this OK?' You'll need to be able to move from inactivity to high-speed action very quickly, without getting bored or complacent when things are quiet. On top of all this, you'll need to be able to communicate well and work within a team.

Does all of that sound impossible to learn? It's not. There are more than 400 operating nuclear power stations in the world, and each one of them has dozens of trained operators. Think of it as being like learning to fly a passenger aircraft—it takes a lot of time (and money) to train a pilot, but there's always one there when you get on a plane. Well, usually.

1.2   What This Book Covers

This is very deliberately not a textbook. It does cover some physics—quite a lot actually, I enjoy physics—but without the maths. In the real world, we usually let the computers do the maths! As a reactor operator, it's the concepts that are going to be relevant to you, i.e. what happens to the reactor, when and why. The book includes more than a hundred diagrams and photos; these should help you understand the more complicated bits. There are a few definitions, but I hope that doesn't put you off; every industry that I know of has 'jargon', and the nuclear industry is no exception. I've included an index of all the essential terms at the back of the book, just in case you need to remind yourself of anything as you go along.

The majority of the reactors in the world, whether generating electricity or powering ships and submarines, are of a particular kind: Pressurised Water Reactors (PWRs), or the similar Boiling Water Reactors (BWRs). This isn't true of the UK where most of the current electricity-generating reactors are of a different type. However, the UK has operated one very successful commercial PWR (Sizewell B) since the 1990s, and is building more e.g. Hinkley

Point 'C', and the reactors planned for Sizewell 'C' and Bradwell 'B'. For this reason, and because that is the author's bias, this book is based around PWR operation and technology.

I'm going to use this book to explain how a PWR reactor works—what makes it run. I'll describe how, if you were a reactor operator, you'd start-up a reactor, change power levels and shut the reactor down. Once you grasp the three key concepts (see below), you'll find that this is all much easier than you might imagine. Along the way, this book is going to introduce you to some of the history of nuclear reactors and power stations. I've always found this interesting, but I also think that it makes it easier to remember the things that affect the operation of the reactor.

I'll explain how a PWR reactor is refuelled and how you tell that the reactor is ready for it. I'm also going to cover several *possible* faults that could happen to a PWR and what you, as a reactor operator, would do about them. Safety comes first, remember? Faults are going to be a big part of an operator's training, however unlikely they may be.

1.3   The Three Key Concepts

Driving a nuclear reactor is not as complicated as you might think, neither is it entirely intuitive. I'm going to suggest that there are three key concepts to understanding the operation of a PWR:

- Reactivity, or how the conditions inside the reactor affect the fission chain reaction.
- Reactor stability, the feedback mechanisms that hold it steady.
- Plant stability, what happens when you connect your reactor to the rest of the plant (and beyond).

If this book helps you to grasp these three key concepts, you'll find it easy to understand the behaviour of a PWR both in its day-to-day operation and during more challenging events.

1.4   And Finally...

If you're thinking of (or have recently started on) a career in the nuclear industry then I wish you good luck, and I hope that you'll find this book useful. If you're just keen on science and engineering, or perhaps you live near to a nuclear reactor, then I hope you'll find this book informative and entertaining.

If you want to read about energy policy and arguments for and against nuclear power stations then find another book; there are plenty out there on the politics of nuclear power. Similarly, this book contains only a brief history of nuclear power stations, and of the significant accidents that have shaped the industry. Once again, there are very many good books on these subjects already written. Instead, this book starts from the *fact* that hundreds of nuclear reactors already exist and are successfully generating electricity. Dozens more are currently under construction and will be running in a few years' time. I'm not going to try to defend these reactors in this book; I'm just going to try to explain how to drive one.

2

Physics Is Phun!

If you're reading this book, I expect it's because you have an interest in science and engineering. That's great, but the problem for me is that I don't know how much you already know. If I guess too low, you're going to feel insulted by what you read. If I guess too high, what I'm saying isn't going to make sense, and you'll lose interest.

So here's the deal: I'm going to start with some physics that I expect you'll clearly remember from school science. I'm going to use that to explain how a nuclear reactor works, and from that starting point, I'll be able to describe how to drive one… feel free to skip anything that you're already familiar with (at your own risk).

2.1   Atoms and Nuclei

You'll probably remember being told that an atom has a (small) positively charged 'nucleus' in the middle, with negatively charged electrons going around it—a bit like planets orbiting the sun. Things are a bit more complicated in the real world, but it's a good enough model for our purposes.

As an example, Fig. 2.1 is an illustration of a helium atom. Helium is one of the simplest of the chemical elements:

Helium has two positively charged particles (called protons, shown in red) in the central bit (the nucleus) and two negatively charged particles (electrons, shown in light blue) going around the outside. The two protons tell a physicist (or a chemist) that this is helium. One proton would make it hydrogen, three would mean it was lithium, four beryllium, and so on through the more

C. Tucker, *How to Drive a Nuclear Reactor*, Springer Praxis Books,
https://doi.org/10.1007/978-3-030-33876-3_2

Fig. 2.1   Helium atom (not to scale)

than a hundred chemical elements that have been found or manufactured. The number of electrons is usually matched to the number of protons, and it's the number of electrons and how they are arranged that determines an element's chemical behaviour.

The two protons in our helium nucleus are positively charged so you'd expect them to push away from each other (like-charges repel, remember?). Because of this, there are also two uncharged particles (called neutrons, shown in dark blue) included in the nucleus to help glue it together. The total number of neutrons and protons is 2 + 2 = 4, so we'd call this helium-4. You can find helium atoms with only one neutron (helium-3), but they are relatively scarce on earth.

Atoms are small. Really small. You could put 100 million of them in a line, and they'd only stretch one centimetre. But atoms are enormous when you compare them to the size of the nucleus. This drawing of a helium-4 atom here is not to scale as the nucleus of a real helium atom is roughly 100,000 times smaller than the size of the whole atom. This book is mostly about what happens in the nuclei of large atoms such as uranium ('nuclei' is the plural of 'nucleus'). It's a book about nuclear physics, not chemistry, so from now on, we'll barely mention electrons. All of the drawings of atoms you see from now on will be of atomic nuclei rather than whole atoms because that's the bit that's interesting to us. If I'm lazy and call these 'atoms', don't worry about it....

So let's look at some nuclei—Fig. 2.2 shows hydrogen-1, helium-4, oxygen-16, iron-56 and uranium-235:

Hydrogen-1 is a single proton with no neutrons. Helium-4 and oxygen-16 have 2 and 8 protons respectively, i.e. they have the same number of neutrons as they have protons. Iron-56 has 26 protons and 30 neutrons (a few more neutrons than protons). By the time you get all the way out to uranium-235 with 92 protons, you find that there are 143 neutrons. This just shows that the

Fig. 2.2   A selection of nuclei

more protons you have in a nucleus, the progressively more neutrons you need to stick it together (this becomes important in Chap. 5 when we talk about radioactive fission products).

By the way, the chemical symbol for uranium is 'U', so I'm going to use 'U-235' instead of uranium-235 from now on; it'll be easier to read.

2.2   Fission

U-235 is not a happy atom… (OK, I mean nucleus, but that doesn't sound as good, does it?)

If you can find a way to give a U-235 nucleus just a bit more energy, it'll probably give up and split into two smaller nuclei. In physics, there are ways of doing this. For a nucleus, a simple way is to drop in another neutron—we could call this process 'neutron capture', if we wanted to be technical about it. The neutron itself won't be carrying a lot of energy, but it will release energy when it joins with the nucleus—think of the noise you get when you snap a magnet onto a block of iron, and you'll get the idea.

That extra energy makes the U-235 nucleus pretty unstable. One way to imagine this is to think of the U-235 nucleus as a large water droplet in a weightless environment such as on a space station. If you've seen any videos of these, you'll have seen how a droplet can start out spherical but if it's poked it can get squashed, stretched and even split into two droplets. If this were to happen, the two new smaller water droplets would probably just sit (float?) there. But that doesn't happen with nuclei as they are each made up of a large number of positively charged protons along with the neutrons—the electrons are a very long way away on this scale and don't really get involved. The two smaller positively charged nuclei will be pushed away from each other very strongly, accelerating to reach tremendous speed before eventually bumping into other atoms and slowing down. Their speed energy will then have been converted into heat. Most of the energy from the splitting of a U-235 atom is carried away by these smaller nuclei.

Fig. 2.3 Fission of U-235

This splitting process is called fission and is shown in Fig. 2.3. The smaller nuclei left behind are usually known as 'fission products'. If a neutron happens to encounter a U-235 nucleus and is travelling slowly enough to be captured, it's then very likely that fission will occur. It's a quick process—for an individual U-235 atom it's all over in around a millionth of a millionth of a second. But the energy released by even a single fission event is enormous on an atomic scale. It's roughly 2.5 million times the energy that you get from 'burning' a carbon atom to make carbon dioxide.

That much energy could probably be quite useful if we could just find a way of encouraging these fission events to happen more often…? Helpfully, U-235 does this for us, because with every fission event we also get two or three spare neutrons. In theory, these could go on to cause more fissions, giving us a 'chain reaction'. In practice, it's a little more complicated.

Most uranium dug up out of the ground (natural uranium) isn't U-235; it's U-238 (it has three more neutrons). It's much less likely that U-238 will fission if it captures a neutron—it's a more stable nucleus. Unfortunately, only around 0.7% of natural Uranium is U-235. You can increase the proportion of U-235 through a process called enrichment, but that's expensive, so most enrichment plants stop at around 4–5% U-235. (If you enrich too far the politics get tricky as you'd effectively be making the raw material for nuclear weapons; let's not go there). That means that you're still stuck with a lot of U-238 in your fuel, which won't fission but can affect the reactor in other ways (as you'll see in later chapters).

At this point, you might be wondering where all this extra energy actually comes from? Here's the trick: if you add-up the masses (weights) of the fission products and neutrons after the fission, you find that they weigh a little less than the U-235, plus the extra neutron, that you started with. The fission process has converted some of the original mass into energy in line with Einstein's famous equation Energy equals Mass times the Speed of Light Squared (E = mc²). It only takes a little bit of lost mass to give a lot of energy as c^2 is such a large number, whatever units you use. Another way to think of it is that the fission products are more tightly stuck together than the original U-235 nucleus because they are smaller nuclei and the attractive forces work better over short distances. In squeezing the nuclei together a bit tighter, a bit of spare energy is released. If you want to know more, go online and look up the physics of 'binding energy'.

2.3   Fast and Slow Neutrons

Neutrons released during and after a fission event are moving at around ten thousand miles per second. Even Physicists are happy to call them 'fast neutrons'. This matters because it makes them far less likely to be captured by a U-235 nucleus. It'd be like shooting a steel ball-bearing past a magnet at very high speed; it's not liable to be stopped. On the other hand, throw a ball bearing slowly towards a magnet, and it'll stop dead. So to encourage further fissions, we need to design a reactor that slows down its neutrons.

An excellent way to slow down fast neutrons is to let them bounce around in some material, giving up a bit of their energy (speed) with each collision. After enough collisions, the fast neutrons will have become 'slow neutrons'. Slow neutrons are sometimes called 'thermal' neutrons. This is because they will be travelling at the same speed as the atoms of the material around them, so to a physicist, they are 'in thermal equilibrium' with it.

Physics has fancy names for this process of slowing-down neutrons: it's called 'moderation', and we call the material in which the neutrons slow-down a 'moderator'. In a 'Pressurised Water Reactor' (PWR) water is used as the moderating material. Physics tells us that more energy is lost in each collision if the atoms that the neutrons collide with are of a similar size (mass) to the neutrons themselves. The hydrogen atoms in water molecules—being individual protons—make water an effective moderator. The other reactors currently in the UK actually use graphite (carbon, another relatively light atom) as a moderator. Chapter 22 briefly covers different designs of reactors, where you can see this.

Incidentally, this is one of the most frequent errors that people make when they talk about nuclear reactors: they'll describe 'control rods' as 'moderating' the reaction. To a reactor operator, the moderator is what makes your reactor run! This error makes me wince every time I hear it, but I'm probably a little over-sensitive…

2.4   Chain Reactions

Once the neutrons released by the fission have been slowed to thermal energy, they are very likely to go on to cause another fission, if they encounter another U-235 atom. But, as you've seen, most of the uranium in a typical reactor is U-238 and, unfortunately, it's pretty good at capturing neutrons as they slow down through intermediate speeds (between fast and slow). In practice, this means that if you simply take uranium and a moderating material and mix them together, nothing will happen. The U-238 will steal all the neutrons before they've had a chance to slow down.

The trick to overcoming this is to use a bit of geometry: deliberately separate the uranium fuel and the moderator. This means that fast neutrons can be produced in the fuel, will escape the fuel into the moderator, where they will then slow down before bouncing back into the fuel, finding a fresh U-235 atom and causing another fission. It all sounds a bit unlikely, but it works! This is the 'chain reaction' that powers your nuclear reactor—it's what makes it a 'reactor'—as you can see in Fig. 2.4. Understanding what affects this chain reaction is the key to understanding the physics of reactors.

At this point, you might be a little bit worried? I said earlier that two or three neutrons are released during or after each fission of U-235 (the average is around 2.4, but you'll never see 0.4 of a neutron). If there's a chain reaction going on and two or three neutrons are produced each time a U-235 atom fissions, then isn't the chain reaction going to snowball very quickly? The answer is 'No'. You're going to waste the majority of your neutrons.

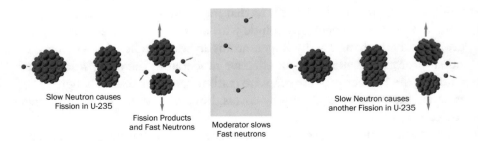

Slow Neutron causes
Fission in U-235

Fission Products
and Fast Neutrons

Moderator slows
Fast neutrons

Slow Neutron causes
another Fission in U-235

Fig. 2.4   The fission chain reaction

Neutrons bounce around at random in the moderator so sometimes fall back into the fuel before they've sufficiently slowed down. If that happens, there's a good chance that they're going to get captured by a U-238 atom and so be lost from the chain reaction. Neutrons are lost in other ways. Some of them are captured by hydrogen-1 nuclei in the water to make hydrogen-2 (also known as heavy hydrogen or deuterium). Others will get captured by the engineering materials, usually metals, which are used to make the structures inside the reactor, such as the cladding that holds the fuel or the 'control rods' (both described in the next chapter). Some neutrons are lost simply because they leak out of the edges of the reactor (an infinite reactor would avoid this, but infinite reactors typically exceed your budget…).

If the reactor is well designed, each fission will produce just enough neutrons for only one of them to go on to cause another fission. A steady number of fissions per second gives a constant rate of energy being released in the reactor. In other words, a stable power level. How many fissions? Well in a full-scale nuclear power station generating, say, 1200 megawatts of electricity, the heat output from the reactor will have to be around 3500 megawatts (1 megawatt is a million watts and is usually abbreviated to MW). The big difference between 3500 and 1200 will be explained in Chap. 10. To produce 3500 MW of heat takes 100 million million million fissions every second. That's a lot of fissions (with no chips).

3

Being Friendly to Neutrons

You can probably already see that you could influence the 'chain reaction' in your Pressurised Water Reactor (PWR):

- You could push control rods into the reactor. Control rods are made of materials that capture (or steal) neutrons. The more they capture, the fewer will be available to sustain a chain reaction.
- A similar effect to using control rods could be achieved by dissolving something in the moderator that captures neutrons. On a PWR this is typically boron, dissolved in the form of boric-acid.
- You could change the temperature of the reactor; this actually produces several different effects so we'll come back to this when we think about stability in later chapters.
- You could also replace some of the fuel with fresh fuel containing more uranium-235, though you'll have to shut down the reactor first!

It's not very useful to only consider whether the changes we are making are helping the chain reaction or hindering it. What we really want is some way to measure the effect. The concept that we use is one of the three key concepts in this book. We call it 'Reactivity'.

C. Tucker, *How to Drive a Nuclear Reactor*, Springer Praxis Books,
https://doi.org/10.1007/978-3-030-33876-3_3

3.1   Introducing Reactivity

Imagine that you are counting all of the fissions taking place in a reactor. By doing this, you will be able to see if the number of fissions taking place (per second, say), goes up, stays the same, or goes down. The number of fissions taking place per second is going to be directly related to the reactor power. If this number is going up, we'll say that the reactor has positive reactivity, and if it's going down, we'll say that it has a negative reactivity. This means that a reactor with a steady number of fissions per second (at constant power) must have a reactivity between the two, i.e. equal to zero. I once met a lecturer in 'reactor physics' from what was then Czechoslovakia. He told me that he taught his students that 'Reactivity is a measure of how friendly a reactor is to neutrons'; which fits quite well with our rough definition.

Reactivity is a nice concept, but how do you make it useful? You'll find that there's very little mathematics in this book, but if you're interested, there a little bit coming-up:

In the paragraph above, I've asked you to think about the number of fissions occurring per second in a reactor. Let's think of this a different way. If we look carefully at a reactor, we could measure the average 'lifetime' of a neutron from the time it is released in a fission event, to the time it is captured by another uranium-235 nucleus, causing a further fission to occur. For reasons I'll explain later, this is quite a long time in nuclear physics terms, at around 1/10th of a second; it's ages on an atomic scale! You can think of this as the time between 'generations' of neutrons (it's all really happening to different neutrons at different times, but the maths still works if you think of it the way I'm suggesting).

Next, we can choose to look at the ratio of the number of neutrons (and hence fissions) in one generation, to the number in the previous generation. If this ratio (usually called 'k') is greater than one, then the number of neutrons in the reactor is rising and so is the reactor power. If it's less than one, then the power is falling. If it's precisely one, the power level is constant. This is similar to our idea of reactivity earlier but, annoyingly, it's centred on one rather than zero.

$$k = \frac{\text{number of neutrons in one generation}}{\text{number of neutrons in the previous generation}}$$

Zero is more convenient as we like the positive/negative definition we started with, so to change 'k' into reactivity, we pull a mathematical trick and define:

$$Reactivity = \frac{k-1}{k}$$

(The division by 'k' is to keep the maths straight, don't worry about it for now).

'k − 1' would also be called 'delta k' by mathematicians (delta being the Greek letter Δ) and reactivity is usually given the Greek letter rho (ρ), so our definition of reactivity becomes the rather neat looking:

$$\rho = \frac{\Delta k}{k}$$

This is better than our first definition because it's quantifiable and measurable. The rate at which the number of neutrons is rising or falling, together with the average neutron lifetime, will give us a numerical value for reactivity rather than just a positive or negative idea. So what sort of numbers are typical for a PWR?

Actually, they are surprisingly small. You'll usually be changing power very slowly in your reactor (except in a fast shutdown or 'trip'), so reactivity can't be moving very far from zero. Put another way, even a change in reactivity as small as 1% (0.01) would be considered a vast change in reactivity on a PWR. In contrast, I was once told that a nuclear weapon might go as high in reactivity as +4% (+0.04), though I've never seen this written down.

3.2 Niles and milliNiles...

Physicists do occasionally make jokes, and there's one to uncover here. The people first working on this subject in the UK got as far as the definition of reactivity written above. Strictly speaking, it is only a ratio so shouldn't have a name (or a 'unit', as we say in physics). The people who dreamt it up didn't like that, so decided to give it one anyway. They chose to call a 1% change in reactivity a 'Nile'. Why? Because a 1% change is a very large delta (k) and what else has a very large delta? *The river Nile.*

I'll be honest. It's not a very good joke, and the Nile isn't an internationally recognised unit. The Navy use something called a 'Derby' and the Americans use 'Dollars/Cents' which are defined differently. The French simply use per cent. However, the civil nuclear industry in the UK is very attached to the Nile, so you'll have to get used to me using it in this book.

Inserting all of the control rods in your PWR might reduce its reactivity by 0.08, or 8 Niles. But in day-to-day operation, reactivity will probably only move by a few thousandths of a Nile (known as milliNiles) away from zero, in either the positive or negative direction. So, in most instances in this book, I'll be using milliNiles when I'm talking about reactivity.

The clever thing about this definition of reactivity (actually because of the division by 'k', I mentioned earlier) is that it allows individual contributions to reactivity to be added-up without any complicated mathematics. So you can work out the reactivity of your fuel (positive) and subtract the reactivity change from inserting control rods (negative). All the other ways in which the chain reaction can be affected can be added and subtracted in the same way to get an overall reactivity for your reactor. Reactor physics is much simpler than it looks!

3.3   Your Reactor's Fuel

This is probably a good time to show you the fuel that goes into your reactor.

Figure 3.1 shows how a 'Fuel Assembly' is put together. There's just under 200 of these fuel assemblies in your reactor's 'Core' and very little else.

On the right-hand side of Fig. 3.1 is a diagram of a fuel pin (or fuel 'rod', the terms are interchangeable). Each fuel pin in your PWR is approximately 12 mm wide and contains 400 or so uranium dioxide pellets. The pellets are roughly 10 mm tall by 10 mm wide, but cylindrical to fit inside the fuel pin. The pellets have slightly dished ends to allow for expansion. At the top of the fuel pin is a space with a spring inside. The spring pushes down on the pellets to stop them moving, and the space created by the spring is somewhere where radioactive gases (from fission products) can accumulate without over-pressurising the pin. The fuel pins themselves are made of an alloy of Zirconium ('Zircaloy') which looks like stainless steel but captures far fewer neutrons than steel. It also has excellent chemical properties meaning that it can with-stand years of operation inside your reactor.

The fuel pins are positioned in a 'Skeleton' that is illustrated on the left of Fig. 3.1. The fuel assembly skeleton consists of a 'Top Nozzle', a 'Bottom

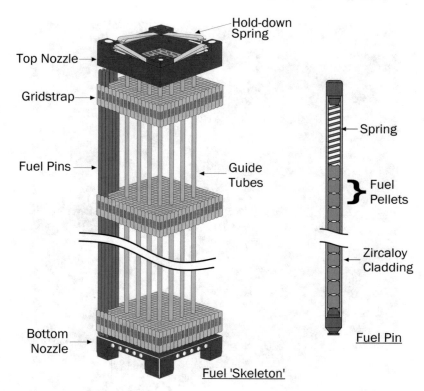

Fig. 3.1 Fuel assembly

Nozzle' and several 'Guide Tubes' (also sometimes called 'Thimble Tubes'), joining them together. 'Gridstraps' are attached to the guide tubes providing holes into which fuel pins are pushed. A fuel assembly arranged like the one in Fig. 3.1 has a 17 × 17 grid of holes in the gridstraps, 25 of which are taken up by guide tubes. This leaves room for 264 individual fuel pins. The fuel pins are packed close together, with just a 3 mm separation at their nearest points. That leaves just enough space between the pins for water to flow. When the reactor is operating the 'Hold-Down Springs' on the top nozzles will be compressed by the weight of the reactor pressure vessel's 'Upper Internals' (see Chap. 6) and so will keep the fuel assemblies firmly in position in the reactor.

Figure 3.2 shows you how one of you fuel assemblies looks before it is loaded into the reactor. It's clearly a long square bundle of fuel pins, but you can also see the gridstraps and the top and bottom nozzles in the picture. Each of these fuel assemblies weighs around 600 kg and costs... more than half a million pounds.

Fig. 3.2 A PWR fuel assembly

3.4 Your Control Rods

I'm going to explain how you'll be using control rods—or not—a bit later in the book. In the meantime, it's worth knowing that PWR control rods are each really a cluster of control rods—we call them 'Rod Cluster Control

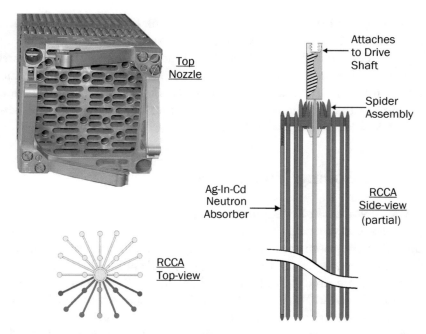

Fig. 3.3 Rod cluster control assembly (RCCA)

Assemblies' (RCCAs). You can imagine an RCCA as being a bit like a 24-legged spider, with the body of the spider being at the top, and each of the legs being nearly 4 m long. This means that an RCCA can slide into the guide tubes of a Fuel Assembly skeleton (which is why they are called 'Guide' tubes), leaving one spare guide tube for other purposes such as power shape measurement.

Figure 3.3 illustrates the RCCA design, and also shows the round holes in the top nozzle of a fuel assembly, each of which leads to a guide tube. If you're observant, you'll also have seen a much smaller hole above the central guide tube, right in the middle of the Top Nozzle. This allows a slight flow of water up the central guide tube. There is a similar hole in the bottom nozzle (not shown) for each of the other guide tubes. If these holes weren't present and the water in the guide tubes was stagnant, it would be heated to boiling by the neutrons bouncing around inside it. It's an essential principle for anyone designing PWRs that you can't have stagnant water anywhere near the core.

The individual rodlets of the RCCA are stainless steel tubes filled with an expensive alloy—80% silver (Ag), 15% indium (In), 5% cadmium (Cd). Why? Because this alloy is especially good at capturing neutrons over a wide range of neutron energies (speeds). This means that you'll get a very rapid shutdown of your reactor if the rods were to fall in. Other control rod materials are available, but Ag-In-Cd is one of the most commonly used in PWRs.

Each RCCA is coupled to a drive shaft which lines up with a penetration in the reactor pressure vessel head (Chap. 6). Above these are hollow tubes and 'Control Rod Drive Mechanisms' (CRDMs). The CRDMs include magnetic grippers that can grab hold of (and move) the driveshafts, and hence move the RCCAs in and out of the reactor.

3.5   The Boiling 'Point' of Water

Apologies if you're already familiar with this little bit of physics; but it's relevant to what comes next…

Go into a classroom (or a pub) and ask people the temperature at which water boils. The chances are that the answer you'll get is '100 °C' (or '212 °F' if any of them are American). Now ask them at what temperature water boils on the top of Everest. You'll probably get the same answer, or you'll hear confusion as to whether the boiling temperature might be higher or lower. The answer is 71 °C, because the air pressure at the top of Everest is just 1/3 of what it is at ground level, and physics says that a lower air pressure lowers the boiling point of water. As another example, astronauts and high altitude pilots sometimes talk about the 'Armstrong Line'—the altitude at which air pressure is low enough for water to boil at normal body temperature. It's around 19 km (60,000 feet) above sea level.

This works the other way around as well. Increase the pressure and the boiling point rises. You could go into a classroom and talk about pressure-cookers if you thought that the kids would have any idea what a pressure-cooker was (they all know about microwaves, but that doesn't help). So, back to the Pub, where people remember such things: in a pressure cooker, the pressure is allowed to rise up to around twice that of normal air. This raises the boiling point of the water inside to about 120 °C, so food cooks more quickly.

I'll just mention that there are a whole set of different units that you could use for measuring pressure. In the UK it's quite common to use 'bar'; where 1 bar is 100,000 Pascals (an SI Unit) and is roughly equal to the normal atmospheric pressure at sea level—it's why the weathermen use millibars in their forecasts. In America, they more typically use pounds per square inch (psi), where 1 bar is 14.5 psi. In France, MegaPascals are used, with 1 MegaPascal

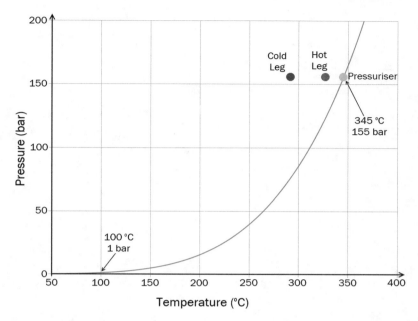

Fig. 3.4 The boiling point of water changing with pressure (saturation curve)

being equal to 10 bar. I'm going to stick with the 'bar' in this book as I think it's a convenient unit for the pressures I'm going to be talking about.

Figure 3.4 shows how the boiling point of water varies with pressure—sometimes called the 'Saturation Curve'. You can see that it boils at 100 °C at 1 bar (atmospheric pressure), and right over the other side of the graph, it boils at 345 °C at 155 bar. 155 bar is the pressure that your PWR 'Primary Circuit' runs at, and this is why it is called a 'Pressurised Water Reactor (PWR)'. You can also probably see how steep the graph is getting at 345 °C—you'd have to go much higher in pressure to drive the boiling point any higher—a challenge for the materials and for the engineers! This is why you won't see many PWRs running at a significantly higher temperature; most are within just a few degrees of each other. It's a shame, as higher temperatures would give better steam conditions, as you'll see later, but you can't change the laws of physics.

I've also marked the temperature and pressure of the 'Pressuriser', 'Hot Leg' and 'Cold Leg' on Fig. 3.4. We'll meet these terms when we talk about your 'Primary Circuit' in Chap. 6, but for the moment, this just gives you an idea of where your PWR will be operating.

4

Criticality Is Not as Bad as It Sounds

4.1 Criticality: One of Science-Fiction's Biggest Mistakes

You'll probably be able to think of a science-fiction film or TV series where one of the characters warns another "The reactor's going to go Critical!" Panic, heroic actions and saving-the-day in the nick-of-time usually follows. There's just one thing: a reactor going critical ('Criticality') isn't a problem at all…

In the last couple of chapters, we looked at how a U-235 atom can be split into two smaller atoms (fission products) releasing energy and two or three fast neutrons. Some of these neutrons might be successfully slowed-down in the moderator and go on to cause fissions in more U-235 atoms. This is the chain reaction that powers your Pressurised Water Reactor (PWR). When running at full power, your reactor undergoes millions of fissions every second.

But what's going on inside a reactor that's shutdown?

4.2 Starting Subcritical: A Shutdown Reactor

Imagine that you've built your PWR. It's hot and pressurised and ready to run; it also has to be at a low enough concentration of boric acid in the moderator, but we'll discuss that later. The only thing stopping it from working is that all of the control rods are fully inserted into the reactor. This means that it has a very negative reactivity. We'll say that the reactor is 'Subcritical'—you'll see why in a minute.

© Springer Nature Switzerland AG 2019
C. Tucker, *How to Drive a Nuclear Reactor*, Springer Praxis Books,
https://doi.org/10.1007/978-3-030-33876-3_4

So, now a bit more physics: it turns out that both U-235 and U-238 atoms occasionally undergo fission without being provoked by a neutron. We call this 'Spontaneous fission', but on an atomic scale, it is pretty infrequent—in a kilogram of U-238, just 5 atoms will do this every second. A kilogram of U-238 contains more than 2 million million million million atoms, so this really is a rare event, and for U-235 it's a thousand times rarer still! This probably seems insignificant, but it isn't. It shows that in your reactor, there will always be a few neutrons being released by natural processes. In fact, many fission products also emit neutrons as part of their radioactive decay (see Chap. 5), so there are going to be even more neutrons flying around in your reactor if it contains anything other than fresh fuel.

What happens to these natural neutrons? Well, they're no different from the ones that we've been talking about when we've been considering the fission chain reaction. They'll have a range of speeds (energies); some fast, some slow. This means that there is always a chance that a neutron will leave the fuel (where it's been produced), be slowed by the moderator, and then find its way back into the fuel to cause fission in another U-235 atom.

This won't be very likely for an individual neutron as all of the control rods are in. Many neutrons will be captured and be unable to cause another fission, but this will sometimes happen, giving short 'chains' of fissions. Any chains of fissions that do occur will each quickly die away—this is consistent with our idea of a reactor being shut down, i.e. 'k' is less than one, reactivity is negative. But, new (short) chains will be being started all the time. So, there are still a few neutrons being produced and moving around—we call it the neutron 'flux'—even in a shutdown reactor. This inevitably means that some neutrons will leak out the edges and you can count them with 'neutron detectors'.

I admit that this may seem at odds with how we defined 'k' (the ratio of the number of neutrons in one generation to that in the previous generation) as that might have led you to expect the number of neutrons to always die away to zero for a shutdown reactor (k less than one). But in a real reactor, the definitions are a little more complicated than the ones in this book. In a shutdown reactor, none of the neutrons from spontaneous fission is going to lead to a self-sustaining chain reaction, but some of them will cause short fission chains that then die out. If you add all of these small chains together at any particular time, you'll have a measurable (low) neutron flux.

4.3   Approaching Criticality

Here's the clever bit: if you start pulling control rods out of the core, you increase its reactivity (reactivity becomes less negative). It's then more likely that an individual neutron will go on to cause another fission. The chains will still die-out, but on average, they will get longer. There will be more active chains at any one time. This is going to increase the neutron flux. It's like pouring water into a cup with lots of small holes, you'll be losing all the water you pour in (through the holes) but the faster you pour the water in, the higher the level of water in the cup will tend to be. In your reactor, the more you pull the control rods out, the less negative the overall reactivity will be, and the higher the neutron flux.

The physics is very helpful here. If you model these effects mathematically, you find that if you halve 'how negative the reactivity is' (say, from minus 8 Niles to minus 4 Niles), then the neutron flux will double. It means you can get a clear idea of what's going on in the reactor, just by watching the neutron counters. You can see this in Fig. 4.1.

Where does this take us? If you keep withdrawing the control rods, you'll eventually reach a point where reactivity is no longer negative. Once reactivity reaches zero, the fission chains (on average) don't die out. This has a particular name: it's called 'Criticality'. Your reactor will have reached a self-sustaining chain reaction. That's all a reactor going 'Critical' means—a reactivity of zero. Practically, you can think of it as 'turning-on' the reactor. The word 'Critical'

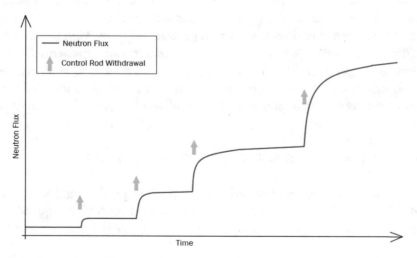

Fig. 4.1   How a subcritical reactor responds to control rod withdrawal

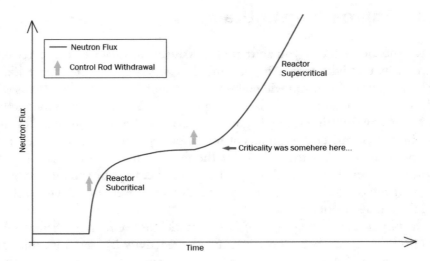

Fig. 4.2 Taking a reactor supercritical

tells you nothing about a reactor's power level, other than that it is steady power; it could be at 1 watt or 3500 megawatts (MW), and it would still be critical. Before we reached criticality, the reactor was 'Subcritical', which just meant that the overall reactivity was negative.

Unfortunately, the closer you get to criticality, the longer it takes for the neutron flux to settle down to a constant value, so it gets progressively harder to see how close you're getting! Physics is sometimes like that. But, if you drive the reactor *through* criticality by pulling the control rods out just a little bit more than you need to, then you'll see something different happen. Rather than the neutron flux detectors showing a gradually steadying neutron flux, you'll see it rise with a very distinctive curve that we call 'exponential'. This is known as 'Supercriticality', and it's illustrated in Fig. 4.2. Overall reactivity will now be positive, and the power you are getting from fission chain reactions is increasing (exponentially, for those of you interested in the maths).

4.4   Supercriticality: Also Not a Problem

What next? Does your reactor suddenly leap up to producing 3000 MW of heat? No. Typically, criticality first occurs with a rate of fissions equivalent to just a few kilowatts of power. Now, bear in mind that the pumps circulating water through your reactor (described in Chap. 6) are adding around 20 MW of heat, and that fission product decay might be responsible for another 10 MW or so. You can see that you're not going to be able to detect the heat (a few kilowatts) from a 'just critical' reactor. But you will be able to see that you've achieved supercriticality because the neutron flux (measured by neutrons leaking out of the reactor) won't settle down to a steady level, but instead, will keep increasing.

To get from a few kilowatts of heat at criticality, up to 3500 MW of heat (for your power station to usefully turn into electricity), you'll need to raise power by roughly a factor of a million. That's going to need the reactor to be supercritical for some time. But don't worry. You'll see in the later chapters on stability, that supercritical does not mean unstable. Once you've reached your intended power level, you can make a small adjustment to reactivity (e.g. by pushing the control rods back into the core a little) and power will steady-out. You'll have returned the reactor to criticality but at a different power level from when you first took it critical.

You can see now what the science fiction writers keep getting wrong, though admittedly, "Captain, the reactor's going critical; but don't worry, that's harmless, and it's what we want to happen anyway to get anything useful out of it…" doesn't have quite the same impact, does it?

4.5   Prompt and Delayed Neutrons

There is one more bit of Physics that's useful here, and that's the distinction between 'Prompt' and 'Delayed' neutrons. In Chap. 2, I told you that each U-235 fission usually produces 2 or 3 neutrons. What I neglected to mention was that they don't all appear at the time of the fission. Most do; these are the 'prompt' neutrons. However, a small fraction (less than 1%) are 'delayed' and are thrown out by some of the fission products a little later. How much 'later' varies a bit, but broadly-speaking, the delayed neutrons can appear anything from a few tenths of a second after fission, out to a few minutes, depending on which fission product they've been released from, and how it decays.

Perhaps surprisingly, the delayed neutrons are really important to designing a reactor. Most practical reactor designs (including PWRs) are built so that criticality can't easily occur on the prompt neutrons alone. Instead, the reactor is intended to only achieve criticality with the delayed neutrons taken into account. Although the delayed neutrons are less than 1% of the total, the delay in their release is very much longer than the lifetime of the prompt neutrons—the time from a fission, through moderation to capture causing a further fission. The delayed neutrons have the effect of raising the *average* neutron lifetime from perhaps a thousandth of a second, up to around a tenth of a second. Relying on the contribution made by the delayed neutrons to achieve criticality slows down power changes in the reactor by about a factor of a hundred. This ensures that reactors are controllable.

If you could add enough positive reactivity to your reactor, you could (in theory) reach a point where it was critical on prompt neutrons alone, i.e. 'Prompt Critical'. Such a reactor would be very difficult to control as power level could move very, very quickly. For a PWR that would mean adding around 0.7 Niles (700 milliNiles) beyond the point where criticality occurs. For reasons that will become clear later in this book, that's really difficult to do. Prompt criticality is easier to achieve in some other reactor designs and was a significant contributor to the Chernobyl event, discussed in Chap. 9.

In the meantime, this is probably a good time to illustrate all of this with a bit of real reactor history.

4.6   Chicago Pile 1 (CP-1)

The fission of Uranium was discovered in 1938 by German Chemist Otto Hahn and his assistant Fritz Strassmann. Not long afterwards, it was realised that a fission chain reaction might be possible. Such a chain reaction had two obvious applications: a compact source of power (electricity) or a powerful weapon. Both of these ideas became significant during and after World War 2, with scientists in Germany and America leading the way in trying to produce a chain reaction. Many books have been written about these projects, but here I just want to focus on one part of the American programme (which you might know as the Manhattan project).

By the beginning of December 1942, in a squash court under Stagg Field at the University of Chicago, 'Chicago Pile 1' was nearing completion. The lead physicist on the project was Enrico Fermi, so CP-1 is sometimes called the 'Fermi Pile'. It was called a 'pile' because, like the experiments that preceded it, it was made up of layers (a pile) of graphite blocks, stacked one on top of another. Graphite is a form of carbon, which is itself mostly carbon-12. Carbon-12 is a light atom and captures very few neutrons, so it's not a bad choice to use as a 'moderator'. What about the fuel? Well, the processes for U-235 enrichment weren't well developed at that time, so the only available fuel for CP-1 was natural uranium, at just 0.7% U-235. The uranium was embedded in holes in the graphite blocks in both metallic form and as an oxide, depending on the supplier.

By the time that the 57th layer of blocks was added, the instruments—counters of neutrons—were suggesting that a controlled chain reaction might be possible, so construction was halted. All told, CP-1 had taken 45,000 blocks of graphite (360 tonnes) containing 5 tonnes of uranium metal and 45 tonnes of uranium oxide. It was 6 m high and more than 7 m across in a roughly spherical shape, supported in a wooden framework.

One of the extraordinary features of the design of CP-1 was that it was built inside a rubber balloon. Air contains nitrogen which was known to capture neutrons so it was thought that it might be necessary to replace the air in the reactor with carbon-dioxide gas—to exclude the nitrogen. As it turned out, CP-1 was built with better quality graphite and uranium than some earlier experiments and the carbon-dioxide was unnecessary, but I can't help being impressed by the fact that the experimenters managed to order a 7.6 m cube-shaped rubber balloon from the Goodyear Company, without telling them what it was for!

Few photographs exist of the construction of CP-1, and I'm not aware of any taken during the experiment described below. However, there are some excellent drawings and paintings of CP-1 by the American artist John Cadel. Figure 4.3 is an example.

You can clearly see the layers of graphite blocks and the (very roughly) spherical shape. The balloon forms a curtain around and on top of the reactor, but it is pulled back like a curtain at the front. Sticking out of the front of the reactor is a primitive control rod. At CP-1 the control rods were made of cadmium strips nailed to flat wooden sheets; as cadmium was known to be effective at capturing neutrons. One of these rods could be dropped in automatically if the signal on the neutron counters went to high. Another was held out by a rope that could be cut with an axe. Still more neutron capturing material was available in the form of buckets of cadmium nitride. The experimenters weren't taking any chances with controllability.

After a false start, when the automatic control rod inserted back into the core (the neutron counting instrument had been set with too low a range), the experiment resumed at 14:00 on the afternoon of the 2nd December 1942.

Fig. 4.3 Chicago Pile 1 (CP-1)

All but one of the control rods was withdrawn, and the final rod was then withdrawn just 6 inches at a time.

I think the event is best described by Physicist Herbert Anderson, who recalled:

> At first you could hear the sound of the neutron counter, clickety-clack, clickety-clack. Then the clicks came more and more rapidly, and after a while they began to merge into a roar; the counter couldn't follow anymore. That was the moment to switch to the chart recorder. But when the switch was made, everyone watched in the sudden silence the mounting deflection of the recorder's pen. It was an awesome silence. Everyone realized the significance of that switch; we were in the high intensity regime and the counters were unable to cope with the situation anymore. Again and again, the scale of the recorder had to be changed to accommodate the neutron intensity which was increasing more and more rapidly. Suddenly Fermi raised his hand. "The pile has gone critical," he announced. No one present had any doubt about it.

Figure 4.4 is that chart recorder trace, with time running from left to right:

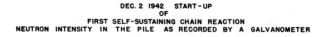

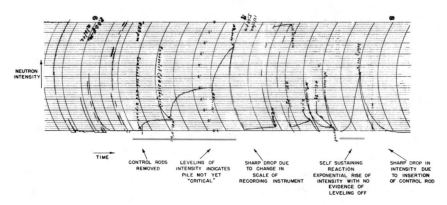

Fig. 4.4 Start-up of Chicago Pile 1

The section I've highlighted in orange clearly shows the way a Subcritical reactor behaves as reactivity is increased. Each time control rods are withdrawn, the neutron flux rises to a new level. The jump up in flux is larger the closer you get to criticality, but it also takes progressively longer to reach a steady value.

In contrast, the section I've highlighted in green shows that criticality has been achieved, with a continuously rising/steepening level of neutron flux. Bear in mind that this part of the trace is actually a much higher flux than that on the left, as the scale of the instrument had been changed by that point. The last part of the trace (after the green highlight) shows the flux dying away as the control rods were re-inserted, the reactor then being Subcritical.

CP-1 had operated (on that first occasion) for 4.5 min and reached a power of 0.5 W. It successfully demonstrated the feasibility of an artificial nuclear reactor (fission chain reaction), and so was the forerunner of all modern reactors, including your PWR. Think of CP-1 as the equivalent of a Newcomen steam engine, but for the nuclear industry. It proved the principles were sound, but the useful applications came later.

Incidentally, some years later, one of the graphite blocks was cut up by the American Nuclear Society to make some rather niche souvenirs. Because of this, I happen to own a very small piece of CP-1 (Fig. 4.5). It's not for sale…

Fig. 4.5   A small (but important) piece of CP-1

5

What Makes Nuclear Special?

There are lots of careers that require years of training, utilise very particular skills, or are focused on helping people when they need it most. None of these reflects the sense in which I'm using the word 'special'. What I do mean is that there are two things to operating a nuclear power station (especially driving a nuclear reactor) that could be considered to be unique to the nuclear industry:

- The compact source of energy
- The heat and the radioactivity from the fission products

In this chapter, I'm going to explain how these make nuclear power special; and why you should care.

5.1 A Compact Source of Energy

Let me recap.

Your Pressurised Water Reactor (PWR) is, in engineering terms, quite small. It is around 4 m tall and just over 3 m across, roughly cylindrical in shape. Its volume is about 35 m³. Bear with me… look around at the room that you're in. If it's 2 m tall and, say, 4 m by 4 m in length and width (about the size of an average lounge), then it has the same volume as your reactor.

Now imagine that your room contains 1 million electric kettles all turned on at the same time. That illustrates how much heat your reactor is producing in the same small space—the 'power density' of your reactor.

© Springer Nature Switzerland AG 2019
C. Tucker, *How to Drive a Nuclear Reactor*, Springer Praxis Books,
https://doi.org/10.1007/978-3-030-33876-3_5

To be honest, you're probably going to struggle to imagine a million kettles. I'm guessing that if you were to pack your 4 m × 4 m × 2 m room full of kettles, you're only going to fit in five to ten thousand of them, depending on their size and shape (and that doesn't allow any space for the power cables). To match your reactor's power output, we're talking 100 times as many kettles in the same space!

Put it another way: if you've got a modern electric kettle, it'll probably have a 3-kW heating element in the bottom. If you put a litre of water in the kettle, it'll bring it to the boil in about 2 min. Your reactor's power is 3500 MW in 35,000 l of volume (1 m³ is 1000 l). This gives it a power density of about 100 kW per litre. At that power density, your kettle would come to the boil in less than 4 s.

When I say that nuclear power stations are a 'compact source of energy', that's what I mean. Power stations are big places, with lots of machinery, pumps, pipes, valves etc. But the nuclear reactor, the thing that drives the whole plant, is tiny by comparison.

This matters. We've said that your PWR generates 1200 MW of electricity from its 3500 MW of heat. The demand on the UK electricity grid varies from say, 20,000 MW on warm summer weekends to around 50,000 MW on cold winter weekdays. It varies quite a bit during each day as well, as you can see in Fig. 5.1. On average, 1200 MW is 3% of demand. So one reactor is satisfying 3% of the country's homes, factories, trains and everything else that's powered by electricity; and it's doing that on just 25 tonnes of uranium fuel per year.

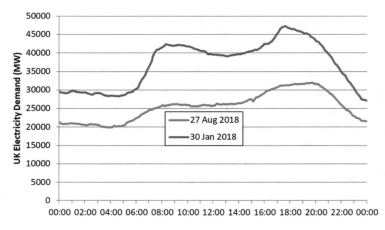

Fig. 5.1   UK electricity demand in summer and winter

Let's say you want to replace the electricity from your PWR with a coal-fired station. How much coal would you need to generate the same amount of electricity? The answer is around 2 million tonnes. That's a pile of coal 150 m high and three times that in width. Remember, that's just for 1 year's generation. The UK is not building any more large coal-fired stations because of their greenhouse gas and acid-rain emissions, but if it was, the sensible place to build them is (and was) next to coal mines, so that you don't have to transport millions of tonnes of coal across the country. In the UK, we still generate more than 40% of our electricity from burning natural gas—but that's easier to transport. However, 25 tonnes of uranium fuel fits on just a few lorries per year. You can build your nuclear power station wherever it's convenient, without having to worry about transporting the fuel.

What about wind power? Well, most large wind turbines are capable of generating about 3 MW of electricity, but the wind is fickle, even in the UK where we have lots of wind! The wind industry will tell you that an average output for an offshore wind turbine is 30% of its rating. So to match 1200 MW, you're going to need to build more than a 1000 large wind turbines. You're also going to want to build them in different places so that they're not all becalmed at the same time—though that might still happen sometimes. I'm not against wind generation—it has a vital role to play as part of a low-carbon energy mix—but it's essential to understand the difference in scale between the output of the few large wind turbines you might see from a beach and the size of the demand on the UK grid.

I said at the start of this book that I'm not going to try to defend the existence of the nuclear power stations; they exist. Hopefully, the paragraphs above will have given you an idea of just how different they are in terms of energy output, compared with other forms of generation. This is one of the things that makes them 'special'. If you're driving a modern PWR at steady output, you'll have as much power at your fingertips as forty fully laden 747 aircraft taking off simultaneously. Just think about that.

5.2 Fission Products

In Chaps. 2 and 3, we looked at the fission process and chain reaction, especially in U-235. In most instances, a U-235 nucleus undergoing fission will split into two pieces, not including any free neutrons that are released. These are the 'fission products'. So what are they?

They're smaller nuclei. You might think that splitting a U-235 nucleus occurs more or less at random, and if you're interested in maths or science,

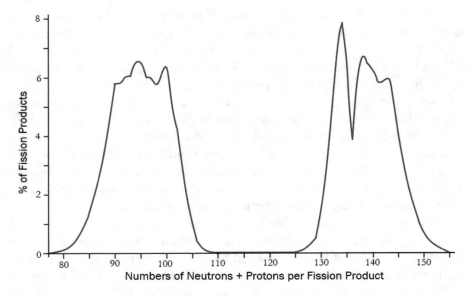

Fig. 5.2 Fission products from U-235 fissions

this might lead you to imagine that you'd see a 'normal' distribution of fission products. It turns out that the internal structure of U-235 is a bit more complicated, and what tends to happen is that one of the fission products will be substantially heavier, one lighter. A near-even split is far less likely to occur than you'd expect. Figure 5.2 shows you the distribution of fission products from U-235 arranged by weight, i.e. the number of neutrons and protons added together. A typical fission would give you one fission product from the left-hand side of this graph, and one from the right.

As you've already seen, being positively charged, fission product nuclei will initially repel each other very strongly. They will leave the fission carrying most of the available energy. After that they will bounce around a bit inside the fuel, slowing down, giving off this energy as heat. Once they've slowed down enough, they'll be able to pick up some stray electrons and become electrically neutral atoms.

But there's a downside. Do you remember that U-235 has 92 protons and 143 neutrons? Nuclei need progressively more neutrons to stay together, the more protons they have. So, having split into two smaller fragments, each is likely to have more neutrons than it needs. In many cases (not quite all), this is going to make the fission products unstable, leading to 'radioactive decay'.

You'd think the obvious way for a fission product to 'decay' if it has too many neutrons, would be to emit a neutron? A few of them do, including those responsible for the delayed neutrons mentioned in Chap. 4. However, nuclear physics is complex, and there are other ways in which fission product decay occurs. You might remember some of this from school science lessons (there are other, more obscure decay mechanisms, but we won't deal with them here):

- Alpha decayis a process that emits a small group of two protons and two neutrons as a fast-moving 'alpha particle'. In reality, this is just a very fast helium nucleus and will slow down pretty quickly. If it is produced within your fuel, it is unlikely to escape it. Emitting an alpha particle doesn't really help with having too many neutrons, so it's an unlikely decay mechanism for a fission product, but alpha particles can sometimes be produced during the fission process itself, e.g. if the U-235 atom splits into three pieces rather than two.
- Much more common for fission products is 'beta decay'. In beta decay, a neutron changes into a proton and an electron. Clearly, this helps reduce the number of neutrons. The electron (beta particle) comes out of the process moving very fast. Beta particles will usually manage to leave the fuel and interact with whatever is outside it. Beta decay also produces exotic-sounding particles called anti-neutrinos. We won't worry about those in this book as they'll leave your reactor (and probably the planet) without interacting with anything.
- Gamma decay occurs either on its own or in combination with alpha or beta decay. Gamma radiation is a higher energy version of an X-ray, so it is very penetrating.

As an example, consider the fission product iodine-131 (I-131). I-131 is produced in approximately 3% of all U-235 fissions—so it's pretty common-place in a running reactor. Iodine 131 decays with a 'half-life' of 8 days—meaning that if you have some I-131 and you wait 8 days, half of the I-131 will have decayed. Wait another 8 days, and half of what you have left will have decayed. And so on…

As you can see in Fig. 5.3, iodine-131 decays by beta decay to xenon-131. Remember that beta decay increases the number of protons by one, so changes the chemical element. It also reduces the number of neutrons by one, keeping the total number of protons and neutrons the same. Shortly afterwards, the xenon-131 nucleus will emit a gamma ray as it'll have too much energy. So I-131 is both a beta and a gamma emitter.

You can get very close to fuel assemblies before they've been used in a reactor as they will be only very mildly radioactive. In contrast, a fuel assembly

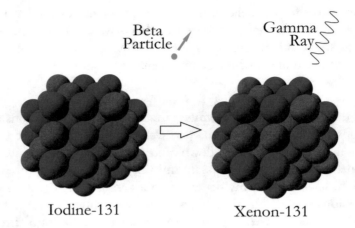

Fig. 5.3 Decay of iodine-131

that's spent up to 5 years producing heat in a reactor ('irradiated fuel') will contain a lot of radioactive fission products. In fact, it will be lethally radioactive. Everything that you do with a fuel assembly after it leaves the reactor must be done with this in mind.

5.3 Decay Heat

We haven't finished with radioactive fission products just yet. They are important to driving a nuclear reactor in another way. We call it 'decay heat'.

Put simply, radioactive decay releases energy. If there's enough radioactive decay, we'll see this energy as heat. Nuclear reactors containing irradiated fuel can produce megawatts of heat, even after they've been shut down. These decays all go on at different rates, and that means that total decay heat falls very quickly immediately after the reactor has shut down, then levels right off. This is because some of the fission products decay very quickly, and others much more slowly. To a mathematician, the decay heat curve is a 'sum of exponentials'.

Figure 5.4 shows a typical decay heat curve:

It's worth thinking about the numbers on this graph. At the instant of reactor shutdown, decay heat is equivalent to about 6.5% of your reactor power. After 1 min that's dropped to 3% and after an hour to about 1%. See how the horizontal scale on the graph is stretched; it starts with seconds and goes all the way out to days. But hang on… If we've started with a reactor running at 3500 MW, 1% power is 35 MW. This graph is telling us that even an hour after shutdown your reactor will be producing 35 MW of heat! After 10 days,

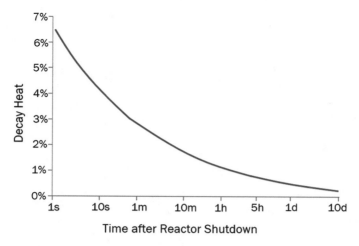

Fig. 5.4 Decay heat as a fraction of reactor power

it will still be producing around 10 MW (0.3% power). If we don't have adequate equipment to remove this heat, the reactor will overheat, and the fuel will be damaged, even though it is already shutdown.

Decay Heat is one of the things that makes nuclear reactors special. In everyday life, when you shut something down, it stops producing heat and just cools down. You can turn off a coal station or a wind turbine and walk away. Radioactive decay of fission products means that nuclear reactors are different, and that's something we'll have to recognise if we drive one.

5.4 The Worst That Could Happen

The world is naturally (and artificially) radioactive. Whether you're thinking of the radioactivity coming from the rocks around you and the food you eat, or the radioactivity left over from the atmospheric nuclear bomb tests in the last century, you live in a radioactive environment.

From the perspective of someone who operates a nuclear power station, the worst thing that could happen would be a large uncontrolled release of radioactive fission products; that is, an event that could significantly increase the risk to the public from radioactivity. This risk only really exists because of the fission products, and that's why I'm mentioning it here. The good news is that the vast majority of the fission products are locked-up within the structure of the fuel so it would take a dramatic event to release any fraction of them.

We'll return to the subject of things that could go wrong with your reactor in a later chapter.

6

The Thing You Put Your Reactor in…

It's time for some engineering… for a Pressurised Water Reactor (PWR) that 'thing you put your reactor in' is called the 'Primary Circuit'. It's sometimes also called the 'Reactor Coolant System' because that's what it does—it contains the cooling water that transports heat away from the reactor.

In this chapter, I'm going to guide you through each of the components of the primary circuit in turn, with diagrams and photographs to help. As a reactor operator, it's going to be essential that you have a good grasp of how this equipment is all put together. It's the only way to really understand how it behaves and what you can do to control it.

Let me start with the layout of the primary circuit, as shown in Fig. 6.1.

This is the PWR cooling loop. To understand it, start at the piece of pipe labelled 'Cold Leg'. On this diagram, the water inside the primary circuit flows from left to right then jumps from the right-hand side back to the left-hand side. I say 'Cold', but this water in this leg is at just over 290 °C. From the cold leg, the water flows into the 'Reactor Pressure Vessel (RPV)'. The water flows downwards to the bottom of the RPV then back up through the 'Core'. The core is where the fuel assemblies are located, so that is where all of the nuclear heat is being produced. The now hotter water leaves the RPV travelling into the 'Hot Leg'. To give you an idea of scale, the hot and cold legs are probably around 0.7 m in diameter—you could crawl along one (I know someone who has!).

In a PWR, the temperature rise of the water as it moves up through the core is surprisingly small, just 30 °C, or so. So the water in the hot leg is probably at about 325 °C. However, 20 tonnes of water flow through the core

© Springer Nature Switzerland AG 2019
C. Tucker, *How to Drive a Nuclear Reactor*, Springer Praxis Books,
https://doi.org/10.1007/978-3-030-33876-3_6

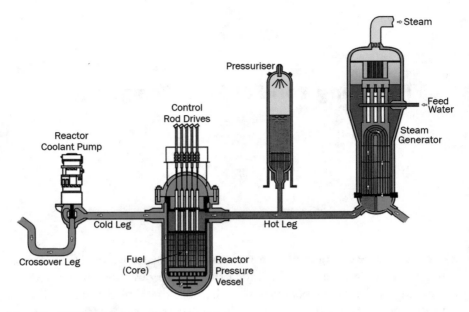

Fig. 6.1 PWR primary circuit cooling loop

every second, so this small temperature rise represents an awful lot of heat energy.

From the hot leg, the water flows up into the 'Steam Generator (SG)'. This contains more than 5000 individual tubes, each like an inverted 'U' in shape. On the outside of the 'U' tubes, there is a separate circuit of water into which the primary circuit water can give up its heat. The function of the steam generator is to remove heat from the primary circuit and use it to boil secondary circuit water into steam. The primary circuit water flowing out of the steam generator will be back at the cold leg temperature (about 290 °C). From the steam generator, the primary circuit water flows into pipework (here called the 'Crossover Leg') that takes it to the 'Reactor Coolant Pump (RCP)'. The electrically driven reactor coolant pump pushes the water back out into the cold leg so that it can travel around the circuit all over again.

You'll see that attached to the hot leg is something called the 'Pressuriser'. I'll come back to this, but first, I need to explain that PWRs are rarely built with just a single cooling loop, as shown in Fig. 6.1. Two, three or four Loops are standard; each connected to the same RPV and with a single pressuriser connected to one of the hot legs. Your PWR is a four-loop plant, as shown in Fig. 6.2:

So, your PWR has one reactor held inside its reactor pressure vessel (RPV). This is connected by four hot legs to the four steam generators (SGs). A cross-

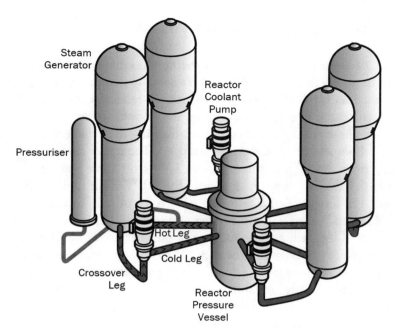

Fig. 6.2 Primary circuit for a four loop PWR

over leg leads from each of the steam generators to the reactor coolant pumps (again, four of these) and then four cold legs connect the RCPs back to the RPV. A single pressuriser is connected by a long pipe to one of the hot legs. It's not the only way of arranging the pipework for a PWR, but it's probably the most common.

The water in the different loops of your primary circuit mixes in the RPV so will be at the same pressure in each of the loops—that's why you only need one pressuriser. Why have more than one loop? Because each loop you add to your design allows you to extract more power (heat) from your reactor. The alternative would be to have fewer loops but have much larger steam generators and pumps—but you'd struggle to transport them from the factory if they got much bigger!

Compared to some other designs of reactor, your PWR primary circuit is quite simple. There are no valves to operate in the primary circuit and no complex pipework. The reactor coolant pumps even run at a fixed speed, giving a (nearly) constant flow of water. There's very little to 'adjust' or 'control' except the reactor itself. But don't worry; you'll see that there'll be enough to keep you occupied as you work through this book.

6.1 The Reactor Pressure Vessel

The reactor pressure vessel (RPV) is the container that holds your nuclear core. In most PWRs (including yours) it's actually two containers in one. The outer container is the RPV-proper, the pressure-retaining vessel that has most of the structural strength. Inside are the 'Lower Internals' and 'Upper Internals'. These are the inner container, forming the structure in which the fuel assemblies and control rods are arranged. The upper and lower Internals are hung from the ledge at the top of the RPV (the 'Core Support').

Figure 6.3 shows a cutaway diagram of the RPV. Your fuel assemblies (the core) are 4 m long, and the complete RPV is 14 m tall. This diagram shows one cold leg on the left and one hot leg on the right. In reality, the number of Hot and cold legs will correspond to the number of cooling loops, so your four-loop PWR will have four of each. The diagram clearly shows the water flow path with water entering from a cold leg and then hitting the side of the lower internals. It's only available flow path is downwards towards the bottom of the RPV, then back up through the fuel. After this, it enters the upper internals and is free to leave into the hot leg.

On Fig. 6.3, you can see that the RPV Head is bolted down so that it can be unbolted and removed for refuelling. The 'Control Rod Drive Mechanisms

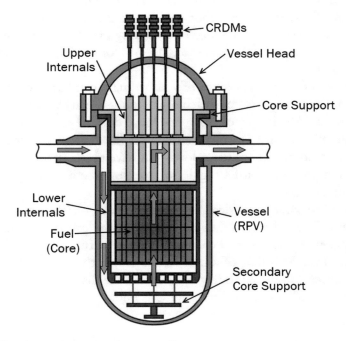

Fig. 6.3 Reactor pressure vessel cross-section

(CRDMs)' are attached to the RPV head with the drive shafts for the control rods being guided by the tubes in the upper internals. At the bottom of your PWR, there is a structure called the 'Secondary Core Support', that's intended to catch the lower internals should the core support fail in a fault. This structure can also be used to guide neutron detectors into the core to routinely measure the reactor's power shape, though not every PWR has this capability.

Figure 6.4 gives you an idea of the scale of a real RPV. This is one being installed at a power station under construction. This RPV weighs 435 tonnes,

Fig. 6.4 A reactor pressure vessel being installed

including the RPV head. It was forged from mild steel sections (for strength) and is internally lined with stainless steel, for chemical protection over its long life.

6.2   The Steam Generators

Apart from the RPV, the largest components in the primary circuit are the steam generators (SGs). In your PWR there is one steam generator in each of the four cooling Loops. This is where the primary and secondary circuits meet, and the secondary side of the SGs will be described in a bit more detail in a later chapter.

After entering the bottom of the SGs from the hot legs, the primary circuit water passes through holes in the 'Tubeplate' and into the 'SG Tubes'. There are more than five thousand thin-walled tubes in each of your SGs. The thin walls allow for the excellent transfer of heat to the secondary circuit water. You can see from Figs. 6.1 and 6.5 that the tubes form a thick bundle of inverted 'U' shapes. In reality, there is still sufficient space between the tubes for secondary circuit water to flow up through the tube bundle, boiling as it does so. Two kinds of mechanical 'Steam Driers' sit above the tube bundle and these will be described in a later chapter. Meanwhile, the now cooler primary circuit water flows out of the steam generators and into the crossover legs.

Figure 6.6 shows a steam generator being delivered to a power station under construction. This model of steam generator is more than 20 m tall and 4.5 m across. An empty steam generator weighs 300 tonnes, so it's not uncommon to see large items of equipment such as SGs being delivered along a river or by sea during power station construction. Road transport of such large items wouldn't be easy.

6.3   The Reactor Coolant Pumps

The reactor coolant pump has a large electric motor, driving a vertical shaft. At the other end of the shaft is a small pump impeller sitting inside a curved pump casing. As the impeller spins, water in the primary circuit is pushed from the crossover leg out into the cold leg—as illustrated in Fig. 6.7.

The motor in your PWR is designed to spin at just under half the speed (frequency) of its electrical supply. So in the UK, where the electricity Grid frequency is 50 Hertz (3000 rpm), this motor is going to turn around 1500 times every minute. Other fractions of Grid frequency (such as a 1/4) are pos-

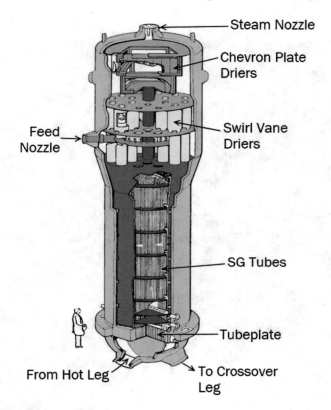

Fig. 6.5 PWR steam generator

sible depending on how the motor is wound. Typically, each RCP will draw 5 MW of power, and they really are large motors, weighing in at about 50 tonnes each, as you can see in Fig. 6.8.

The shaft transfers the speed of the motor down to the impeller (a bit like a propeller, it's the bit that pushes the water). But here's the challenge… the impeller is inside the primary circuit, so it is sitting in water at 155 bar and 290 °C. The motor is inside the 'Reactor Building' but outside the primary circuit. That means that the spinning shaft passes through the primary circuit's pressure boundary. How are you going to seal it so that primary circuit water doesn't come out around the shaft? This is difficult as there's no simple mechanical sealing arrangement that's going to do the job. Instead, the technique that your PWR, and most others, have adopted is to inject clean water at a higher pressure than 155 bar into the sealing package. Some of this water flows down into the primary circuit, and some flows back up the shaft to be collected and returned later. In other words, the RCPs rely on 'Seal Injection'

Fig. 6.6   Delivery of a PWR steam generator

to maintain their seals and keep the primary circuit water inside the primary circuit.

6.4   The Pressuriser

The pressuriser is used to control the pressure in the primary circuit. It's what gets it to 155 bar, and keeps it there. On your PWR the pressuriser is a sizeable tube-shaped tank, mounted vertically. It's got around 2 MW of electric heaters in the bottom and a few smaller connections at the top. If you turn the heaters on—something you can do from the control room—you'll heat-up the water inside. Once this starts boiling you'll get a bubble of steam in the top. The more steam you get, the higher the pressure will rise (as it pushes

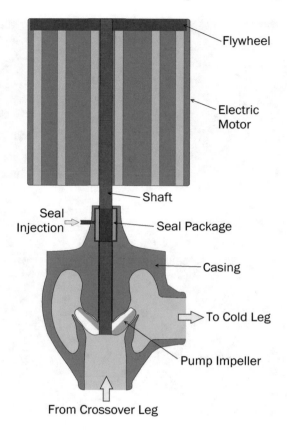

Fig. 6.7 Sketch of a reactor coolant pump

down on the water) and that will shift the boiling point of the water along the saturation curve that you'll remember from Chap. 3. Eventually, the pressuriser will reach 345 °C and 155 bar. You can turn the heaters right down when you get there; as that's the pressure you want to be at when running your PWR.

The bottom of the pressuriser is connected with a long piece of pipe to one of the hot legs. Through this pipe (the pressuriser 'Surge Line'), the pressure from the pressuriser steam bubble is shared with the rest of the primary circuit. The surge line is also a route for primary circuit water to enter or leave the pressuriser as the water expands and contracts with temperature. The long line is used to reduce temperature cycling at the bottom of the pressuriser; as that could shorten its life through fatigue.

Fig. 6.8   A reactor coolant pump (RCP) motor

If at any time you want to raise the pressure in the primary circuit, just increase the pressuriser heating power. If you want to lower the pressure, turn the heaters down a bit. If you need to drop the pressure more quickly, you can open a valve that lets colder water flow through spray nozzles at the top of the pressuriser. Designs vary, but on your PWR this spray water comes from one of the cold legs, so it's more than 50 °C colder the steam bubble. If you use the sprays, it'll cause some of the steam to condense, and the pressure will drop quickly. It's also quite common to fit 'Pressure Relief Valves' to the top of the pressuriser. These valves protect the primary circuit from overpressurisation if the sprays can't cope or can't be used.

Figure 6.9 shows a cutaway diagram of your pressuriser, clearly showing the electric heaters, the surge line connection and the spray nozzles. This example is more than 15 m tall and weighs nearly 100 tonnes even when empty.

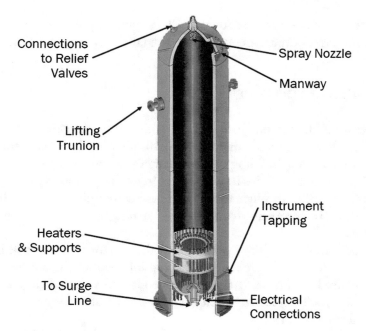

Connections to Relief Valves

Spray Nozzle

Manway

Lifting Trunion

Instrument Tapping

Heaters & Supports

To Surge Line

Electrical Connections

Fig. 6.9 The pressuriser

6.5   Putting It All Together

It's worth reflecting on how much engineering goes into designing, manufacturing and assembling the components of the primary circuit. It needs to be made of very high-quality materials, rigorously forged and welded together with very many quality checks along the way. The various components then need to be delivered to your construction site and welded together within your reactor building, before testing and commissioning of your plant can begin.

Figure 6.10 shows an example. In the picture, you should be able to identify the four steam generators covered in mirror-finish insulation to reduce heat losses. If you look up from the tops of the steam generators, you'll be able to see the main steam lines which are heavily buttressed against steam forces and earthquakes. Between the steam generators is the 'Refuelling Cavity' and in this, in day-to-day operation, sits the RPV head with more mirror-finish insulation covering its base. The reactor pressure vessel itself, together with the core, are below the level of the refuelling cavity floor so can't be seen in this picture. The reactor coolant pumps are hidden beneath the concrete floors on either side (here boarded out for construction work). The big cooling fans (the light grey cones) are directly above the RCPs. Finally, you can just see the top of the pressuriser in the concrete box on the far left… and lots of cranes!

6.6   Inside the 'Can'

The reactor building is 'the thing you put your primary circuit in'. Those built for modern PWRs (like the one in Fig. 6.10) are both large and strong. The walls of your reactor building will be made of concrete, more than a metre thick. The concrete will be heavily re-enforced and made even stronger by steel cables running through tubes in the concrete pulled into tension after the building is complete—concrete gets stronger when it's compressed in this way.

It doesn't stop there. The inside wall of the reactor building will be lined with welded steel. Every pipe or cable going in or out of the building will be welded or fixed to the steel liner. When people tell you that they are 'going into the can' you'll understand that they say this because they are visiting the steel-lined reactor building.

The combination of concrete walls and steel liner mean that the building is immensely strong and also air-tight. You could pressurise the whole building

Fig. 6.10   The primary circuit installed in a reactor building

up to more than 3 bar, with steam from a broken pipe, say, and it won't leak. The building will also be able to withstand impacts from outside and cope with earthquakes. This is why reactor buildings are sometimes called 'Containment' buildings. Nevertheless, it's not uncommon for modern PWRs to have a second reactor building built around the first one, just in case the inner one suffers from leaks. In these designs you have both a 'Primary Containment' and a 'Secondary Containment' building, though, perhaps confusingly, these names have nothing to do with the primary and secondary circuits of your PWR.

6.7   A Sense of Scale

- Your primary circuit is operating at 300 °C and 155 bar.
- Your reactor is producing 3500 MW of heat.
- The water flow around your primary circuit is around 20 tonnes per second.
- Your reactor coolant pumps are spinning 1500 times every minute, driven by 50-tonne electric motors.
- And it will operate reliably for 1–2 years at a time, without a break.

That's why you need engineers!

7

Pull the Rods Out and Stand Back

Now that you understand what powers your PWR—both in terms of the physics and the engineering—it's probably time to get behind the controls and have a go at driving it. You're going to begin with a reactor start-up; think of it as pulling away from the curb…

7.1 Where Do You Start?

You might imagine that we'd start this chapter with the reactor cold and depressurised. In fact, you're going to begin the reactor start-up procedure with your reactor already sat at 'Normal Operating Pressure' and 'Normal Operating Temperature' (NOP/NOT). As a reminder, the normal operating pressure is around 155 bar, and normal operating temperature is with a cold leg temperature (known as 'Tcold') of about 290 °C. With the reactor shut-down, the temperature in the hot legs ('Thot') will be very close to Tcold.

There are a couple of reasons for starting this chapter at NOP/NOT. Firstly, if your reactor has just shutdown in an unplanned way, then NOP/NOT is where your control systems will have stabilised the plant. This will have avoided an unplanned cooldown or a loss of pressure that could have led to boiling in the core. Secondly, to start-up the reactor, you'll need supplies to your control rods. You'll only have these if the plant hasn't 'tripped'. As with many other PWRs, you will have a trip if your primary circuit temperature or pressure is outside of the normal range for a running reactor. In other words, it's not possible to start-up your reactor (at least not using standard procedures) unless you're starting from pretty close to NOP/NOT. We'll talk about

© Springer Nature Switzerland AG 2019
C. Tucker, *How to Drive a Nuclear Reactor*, Springer Praxis Books,
https://doi.org/10.1007/978-3-030-33876-3_7

moving the plant from a cold state to NOP/NOT when we discuss refuelling shutdowns in Chap. 21.

The conditions for starting-up the plant are even more restrictive than you might imagine. At your plant, to keep things simple, it's not permitted to run the reactor with anything less than all four cooling loops in service. This means that you need to have all four reactor coolant pumps running and a practical water level in each steam generator. You'll need a way of removing heat from the primary circuit—both decay heat and heat added by the reactor coolant pumps—and that will involve dumping steam. Typically, you'll be dumping steam to the 'Turbine Condensers' (see Chap. 12), and that will require you to be running the large pumps that circulate seawater through the condenser tubing.

Chemistry control is going to be vital so as to avoid damaging the plant, so you may well already be running 'Main Feedwater Pumps' (also in Chap. 10); not to feed the steam generators, but instead to circulate secondary circuit water through chemical clean-up systems. Added to this you'll have the usual computers, instrumentation, Heating, Ventilation Air Conditioning (HVAC), lighting and plant cooling systems in service.

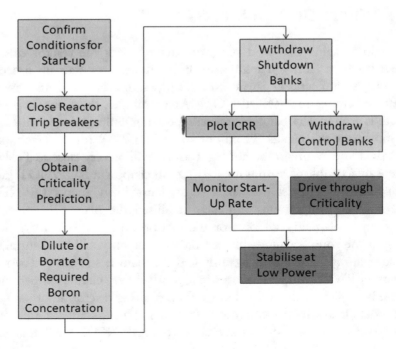

Fig. 7.1 Start-up flowchart

To give you an idea, to run all of this equipment, might take 40 MW of electricity from the electricity grid, that's before you've even begun your reactor start-up! This is one of the reasons why very few PWRs can start-up without grid supplies, i.e. they can't perform what's known as a 'Black Start-up'. Some other power stations will have to be generating first, to give the electricity to run all of your plant systems…

An overview of your Start-up Procedure is shown in the flowchart in Fig. 7.1. You can see that the first step in your procedure is the one that ensures that you meet all of the requirements I've described above—and probably a few others. Let's assume that goes well, and move to the next step.

7.2   Are You Protected?

You'll be starting-up from a 'tripped' state. This means that the circuit breakers between the control rod drive mechanisms (CRDMs) and their power supplies will all be open. This stops you from moving any control rods so you'll clearly need to close these breakers before you can start-up. In fact, to close the breakers you'll also need to reset any 'trips' that your 'Reactor Protection System' has applied otherwise the breakers won't close, but don't worry about that for now; we'll talk about protection systems in more detail in Chap. 17.

In terms of nuclear safety, closing the circuit breakers gives you more than just the ability to move the control rods. It also means that the protection system can automatically shut down the reactor at any time by re-opening the circuit breakers and dropping the control rods back in. Closing the breakers 'arms' the control-rod-tripping part of the protection system, making it a bit safer to do other things such as changing the boron concentration in the primary circuit.

7.3   Predicting Criticality

One of the things that make PWRs different from some other kinds of reactors is that there are two ways in which the operator can intentionally change the reactivity: control rods and boron. Arguably, changing temperature is a third way of adjusting reactivity (as you'll see later), but it's not really relevant here, as we've said you're starting from stable conditions at NOP/NOT.

The higher the boron concentration, the less reactivity the core will have. This would mean that you'd have to pull the control rods out further to achieve criticality. Conversely, if you reduce the boron concentration, you'll be able to go critical with the control rods more deeply inserted. If you move the boron far enough, your PWR might go critical with all of the control rods fully inserted; or not go critical at all, with all the control rods fully withdrawn! Clearly, you're going to want to think about this carefully before you begin to change anything.

In a later chapter, you'll see all of the steps that go into predicting criticality, i.e. the combination of boron concentration and rod withdrawal that will give you the criticality that you want, when you want it. For now, just accept that you'll be given a prediction by someone who has already done the calculations. Once you have that, you're ready to move on.

7.4   Changing Boron

There are more than 250 tonnes of water circulating around your primary circuit, so you might be wondering how you go about changing its boron concentration? The answer is to use a system called the Chemical and Volume Control System (CVCS). There's a diagram of your CVCS in Fig. 7.2, with the flow of water running clockwise from the top left.

Let's start with the 'Letdown Orifices'. These are connected to one of the crossover legs on your PWR, and they provide a high-resistance flow path for

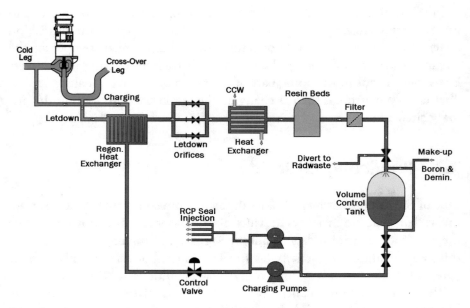

Fig. 7.2 Chemical and volume control system (CVCS)

water leaving the primary circuit. Each 'orifice' is actually a plate with many small holes drilled in it. It takes a lot of pressure to drive the water through these holes, and that in turn causes a pressure drop, so that the water downstream of the orifices is at a much lower pressure than that on the primary circuit side.

If we simply dropped the pressure through the letdown orifices, we'd see the water flashing off to steam as it would be above the boiling point for the new (lower) pressure. Because of this, we also need to cool the water that is being 'let-down' into the CVCS. This begins upstream of the orifices in the 'Regenerative Heat Exchanger'—with colder water being 'charged' back to the primary circuit being used to cool the water that's letdown. This is followed by a more conventional 'Non-Regenerative' heat exchanger cooled by a separate cooling system (the 'Component Cooling Water System', see Chap. 21). Once the water is cooled and depressurised, we can use filters and chemical resin beds (a fine filter made-up of chemically coated plastic beads) to clean up the primary circuit water and remove any impurities that have dissolved in it, for example, from corrosion.

The water falls into a large tank (the 'Volume Control Tank (VCT)') where radioactive gases can be removed. Hydrogen is present within this tank, keeping the oxygen level in the water very low to reduce corrosion of the primary circuit. You'll also see connections here to the 'Reactor Make-up Water System'. This is where you can add pure water or borated water (water with boron in it). Each of these is kept in tanks and can be pumped into the CVCS, separately, or as a 'blended mix' if you're trying to add water at a specific boron concentration. Adding fresh water 'dilutes' the primary circuit. Adding boron 'borates' it. You'll also see these operations described as performing 'dilutions' and 'borations'.

Obviously, you can't just keep adding water to the primary circuit as sooner or later you'll have too much water! The other feature you can see on the figure is a path off to your radioactive waste plant. This is how the CVCS is used to 'divert' excess water, thereby removing it from the primary circuit.

Finally, you'll see that the water leaves the VCT and enters the 'Charging Pumps'. These are very high pressure pumps (discharging at around 190 bar) and are used to push the water back into one of the cold legs on the primary circuit. They also provide the reactor coolant pump seal injection flow that we met in the last chapter. The 'Charging Flow Control Valve' downstream of the charging pumps provides the overall control for the amount of water going back into in the primary circuit. If this valve opens ups, the pressuriser water level will rise and the VCT level will fall; and vice versa if the valve closes-in.

Let's put some numbers on this: as you've already seen the primary circuit flow is around 20 tonnes per second, i.e. that's how much water flows through the core every second. In contrast, the CVCS flow is approximately 20 tonnes per hour, so is thousands of times less. Using the CVCS, it's possible to increase or decrease the primary circuit boron concentration by a few 100 ppm (parts per million) within an hour or so. That doesn't sound a lot, but typically, when running at full power, you'll only need to change the boron by 2 or 3 ppm in a whole day.

A reactor start-up is a bit different from day-to-day operations. You may well find that you'll want to change the boron by a few 100 ppm to match the criticality prediction you've been given. You do this before you start to move control rods, so that you're only affecting reactivity in one way at a time, with less chance of getting it wrong. You'll then need to be prepared to change the boron concentration by hundreds of ppm as you raise power. All of this is going to take a few hours, and your plans will have to allow for that.

You could do these things in a different order, but on this particular start-up, I'm going to assume that you'll use the CVCS to achieve your target boron concentration before moving any of the control rods. It does also work the other way around, as you'll see in a later chapter.

7.5   First Steps

You're not usually going to move rods individually—there's more than 50 so that would take a while. Instead, the rods are divided into 'Banks' each containing several rods, arranged symmetrically in the core. You'll only plan to move whole banks at a time. This will help to keep the power in the reactor symmetric, i.e. you won't have all the power on just one side of the reactor.

Although all of your control rods are identical, for practical purposes, they are divided into two sets—the 'Shutdown Banks' and the 'Control Banks'. The shutdown banks are either fully in—when your reactor is shutdown—or fully out (when it is running). They play no part in controlling your reactor other than in shutting it down and keeping it shut down. It's the control banks that are the groups of rods that are used to control reactivity as you approach and drive through criticality. When the reactor is running at full power you'll have very few rods inserted, in your case less than half a dozen from the control banks; and they'll only be inserted 20 cm or so into the top of the reactor. We'll explain why this is, a bit later.

Control Rod Drive Mechanisms (CRDMs) move the control rods in steps—these relate to the spacing of the rings on the control rod drive shafts (the bits that the grippers grab hold of). Typically, these steps might be 15 mm or so apart, so there'll be more than 200 steps separating a rod that's fully inserted into the core, from one that is fully withdrawn.

On your particular PWR, there are half a dozen shutdown banks and three control banks. There's a selector switch for you to choose each bank in turn, but the control rods are actually moved by you pushing or pulling a simple joystick (Fig. 7.3). Driving a nuclear reactor can sometimes be that easy!

Having achieved the appropriate boron concentration, you're going to move the shutdown banks, one at a time, from fully inserted to fully withdrawn. The CRDMs allow you to do this at around 50 steps per minute, so to fully withdraw a bank is about 5 min work. Withdrawing your six shutdown banks ought to only take 30 min, or so. Seems fast? Well, remember that the core is only 4 m tall, so you're not moving the rods very far.

In your control room, you're going to want to move things a little slower than this: What if your prediction of criticality were wrong? Or you had an incorrect boron concentration indicated on your instruments? It makes sense to monitor your reactor carefully as you withdraw each shutdown bank, just to be sure that you don't reach criticality unexpectedly. How do you 'monitor' the reactor? By watching what happens to the number of neutrons leaking out

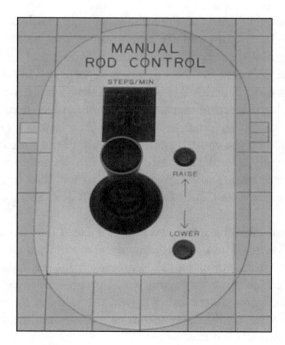

Fig. 7.3   Control rod joystick

of the edges of the reactor (the neutron flux) as the rods are withdrawn. The more neutrons that are flying around in the reactor, the more will leak out of the edges and reach your neutron counters (flux instruments).

7.6   Approach to Criticality

OK, now the significant bit. On the one hand, I've told you that criticality is not something to be afraid of; that it is the normal state of a running reactor. Now I'm going to say that despite this, it's something that you need to approach very cautiously. The problem is that moving the control rods can add a lot of positive reactivity very quickly. If your prediction of the estimated point of criticality is wrong, or you're not paying attention, you could take the reactor significantly supercritical before you realise what's happening. If the reactor is very supercritical, the rate that the power would be increasing would be too high, and you'd be relying on the protection system to automatically shut down the reactor. At best, you'd have to begin your start-up all over again. At worst you might have damaged some of the fuel.

Your problem is how neutron flux rises as you approach criticality. There's no visible line that you'll cross when you get there. The level of flux you measure at criticality will depend on the burn-up of the fuel, the boron concentration, control rod insertions and the calibration of your neutron counters. So how can you cautiously approach something you can't see?

Do you remember that in Chap. 3, when we were looking at how the neutron flux changes in a subcritical reactor, I said that if you halve 'how negative the reactivity is', the neutron counter signal will double? It turns out we can use this to our advantage. What we're going to do is to plot a graph of how 1/ counts (1 divided by the counts) changes as we pull the control banks out of the reactor. Actually, to make the vertical scale simpler, let's plot initial counts divided by current counts as we go along, as that will start from a value of 1. This is called the 'Inverse Count Rate Ratio' or ICRR. You can see an ICRR plot in Fig. 7.4.

An ICRR plot always starts at a value of 1 (seen here on the right-hand scale). As the Control Banks are withdrawn the reactivity increases as does the neutron flux, with the count rate shown here on the left-hand scale. As the distance from criticality reduces, the ICRR graph falls towards zero. More usefully, as the distance to criticality halves, the value of the ICRR will also halve (as the flux has doubled). If you draw a line extending from the graph down to the x-axis (the control rod withdrawal), this will give you a prediction of criticality before you get there. So by plotting an ICRR graph as you

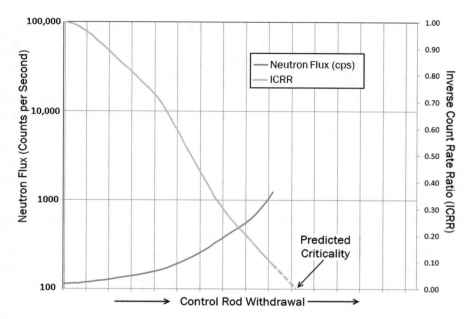

Fig. 7.4 Start-up flux and inverse count rate ratio

withdraw the control banks, you can ensure that you don't blindly go through the point of criticality.

You might also notice that the ICRR plot isn't a straight line; it curves quite markedly at the start (left-hand side) of the graph. This is because the amount of reactivity added to the core by each step of control rod movement is not constant. Less reactivity is added when the rods are only slightly (or very deeply) inserted than when they are in the middle of the core, and that's what gives the ICRR plot this characteristic shape.

7.7 Waiting for Criticality…

One of the most common errors that new reactor operators make—though usually in a simulator—is to try to balance the reactor precisely at the point of initial criticality. Why is this an error? Because it takes forever…! You might also remember from Chap. 3 that the closer you get to criticality, the longer it takes for the neutron flux to settle down to a constant value, so to achieve this so you'd have to wait longer and longer to be sure that you haven't yet reached it. It's pointless, but it's easy to waste an hour or two doing this. It's dull.

On the other hand, if the ICRR plot is showing you approximately where criticality lies, why not just step the control rods out a little bit more so that you know you're just past this point. You should then be able to see a persistently rising flux in your (now) slightly supercritical reactor. At such low power, a small amount of positive reactivity is not a problem, and you can stop the power rise at any point just by driving the control rods back in a few steps.

What would be the desired stabilisation point? A low flux level, but one at which you're sure could only be achieved on a critical reactor. I'm going to get you to stop the power rise at one ten-thousandth of one percent power. That's around 3 kW of nuclear power, so it's a minimal heat input compared with the decay heat and the heat from the RCPs. It's a chance to look around at all of your indications and check nothing is amiss before going any further. In particular, it's a chance to confirm how accurate your criticality prediction was. If criticality occurred a long way away from your prediction that could indicate a bigger problem. Perhaps you've left a control rod behind, or maybe you have an issue with your boron measurements? You'd definitely want to investigate before going any further.

7.8   Doubling Time and Start-Up Rate

Now that your reactor is critical at low power, you're going to want to increase power up to a more useful level. This is easy. Pull the control rods out a few steps. The reactor will be supercritical, so power will rise exponentially. How fast the reactor power increases will depend on how far supercritical the reactor is. So how far supercritical should it be and how do we measure the rate of increase?

One traditional way of measuring how fast the power is rising in a reactor is using 'doubling time'. It's sounds a simple concept: the doubling time is just the time it takes for the reactor power to double. While it's simple, it's not very easy to use. The doubling time approaches zero for a reactor where the power is rising very quickly, and it's infinitely long for a reactor with very stable power. Even worse, a falling power level will have a negative doubling time—approaching negative infinity as it stabilises. I find it hard to visualise…

PWRs typically use a more straightforward measure of how reactor power is changing; it's called the 'Start-Up Rate (SUR)', and it's measured in units of 'Decades Per Minute' (DPM). A reactor that is at stable power has a zero value of SUR. If the power is rising by a factor of 10 every minute, it's said to have a SUR of plus one DPM. If power is falling by a factor of 10 every minute, its

SUR will be minus one DPM. A SUR of plus one (with power rising by a factor of 10 every minute) would be a ridiculous place to be for a reactor running at full power. But if you're starting at a ten-thousandth of one percent reactor power, a SUR of plus one means that your reactor would take 4 whole minutes to even reach 1% power. That's plenty of time to step the control rods back in. A limit of plus one DPM is what we'll use now as we raise reactor power towards a few percent of full power.

So, pull the rods out (a few steps) and stand-back.

7.9   Where to Next...?

In this chapter, you've been starting-up your reactor. You've driven it up to a few percent power (say, a couple of 100 MW). From initial criticality and very low power, you'll have achieved this by withdrawing control rods to give a positive SUR.

But...as power rises through a few percent of full power you'll start to see other things happen on your plant:

- Thot will begin to increase above Tcold—perhaps that isn't so strange? Your reactor is now producing many megawatts of heat which is being transferred to the cooling water in the primary circuit.
- The rate at which you're dumping steam will increase, and you'll need to increase feedwater flow to maintain steam generator water levels; again there's more heat to get rid of, so this isn't surprising.

But then there's these two:

- The level of water in the pressuriser will rise
- The SUR will drop back towards zero, without you moving the control rods.

All of these effects are due to reaching what we'll call 'the point of adding heat,' i.e. they are related to temperature. They actually make it easy to stabilise the reactor at just a few percent power. We'll come back to the these effects in later chapters. But first, we need to think about how we're going to measure reactor power.

8

Watt Power?

Knowing your reactor's power level is going to be essential; much as you need to know the speed of any car that you drive. In a car, the speedometer works by counting how many times the wheels go around, but how do you measure the power of your reactor?

You'll remember that as you watched your reactor approach criticality, you could estimate its reactivity by counting the neutrons that were leaking out of the sides of the reactor. Is there anything stopping us from using this technique to measure the reactor power?

In theory, no. The higher the reactor power, the more the neutrons will leak out of the sides. There's a practical problem, though. The speedometer in your car could measure your speed from a few miles per hour up to, say, a couple of 100 miles per hour. It might not be very good at measuring 1 mph, or 1/10th of a mile an hour? I'm guessing that it would utterly fail to measure 1000 mph? So its highest accurate speed measurement is about 100 times its lowest accurate speed measurement. A measurement range of times a hundred. That's good enough for a car.

What about your reactor? We saw that it went critical at just a few kilowatts of power, but at full power, it'll be running at 3500 megawatts (MW). That means that the highest power is roughly a million times the lowest power we want to measure. If neutron leakage rises in proportion to power level that tells us that we'll be looking to measure neutron leakage over a similar 'times a million' range. I'll be honest. No single instrument is going to do that with any sensible accuracy.

There's a simple answer to this problem—and it's quite common across reactors worldwide. Install more than one instrument and switch between

© Springer Nature Switzerland AG 2019
C. Tucker, *How to Drive a Nuclear Reactor*, Springer Praxis Books,
https://doi.org/10.1007/978-3-030-33876-3_8

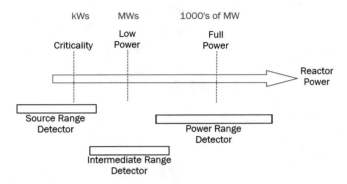

Fig. 8.1 Flux instrument ranges

them. In the case of neutron leakage, we can arrange to have some very sensitive instruments able to measure the low leakage as your reactor is taken critical. We can have another set of instruments that measure from criticality up to part power (perhaps a few tens of percent power) and a final set of instruments that measure from there all the way up to full power, and a bit beyond. This is illustrated in Fig. 8.1.

You'll notice how the ranges of the instruments overlap. That way, no matter the power level, you'll always have at least one set of accurate instruments to measure your reactor power. Reactor operators tend to refer to these instruments as 'Flux' measurements as they measure the number of neutrons that are leaking from the reactor (you'll remember we used the same word for the number of neutrons moving around inside the reactor). Common terms for the three ranges of instruments are 'Source Range' for the lowest, 'Power Range' for the highest and 'Intermediate Range' for the one in between. The number of ranges and what they are called might vary from plant to plant, but the principle is the same.

8.1   Three Problems with Flux

So, that's it, problem solved. We'll simply use neutron flux instruments to measure the reactor power? Unfortunately, there are a few problems….

Neutron flux instruments are great things to use on a minute-to-minute, hour-to-hour basis. They'll respond very quickly to reactor power changes because neutrons travel very fast. Modern instruments are also very reliable, which is good because it wouldn't be straightforward to change one while it's installed next to a running reactor!

But they drift… by which I mean that in a few days or so they will slowly have lost accuracy and they'll be telling you an incorrect power level. This isn't because there's something fundamentally wrong with the instruments; it's because of changes that are going on inside the reactor.

After you've loaded a reactor with fresh fuel, or a mixture of fresh and irradiated fuel, the chances are that the power shape is going to be concentrated towards the middle of the core. This is because neutrons produced near the edges are likely to leak out of the core so there'll be a natural tendency for the number of neutrons and hence the power level (number of fissions) to be lower towards the edges of the core than in the centre. I've tried to show this in Fig. 8.2.

Remember that your flux instruments measure leakage from the edges of the reactor—neutrons don't travel very far in water so these instruments can't 'see' the centre. This means that the power level they are recording is from quite a low level of neutron leakage. Strictly, I include the top and bottom of the core when I talk about 'edges', but you probably won't have any flux instruments in those directions, so we could just say 'sides' instead.

Now think about what's happening in the fuel. As fissions take place, you will be using up U-235 atoms and building-up fission products, some of which capture neutrons. This will be happening faster in the centre of the core because that's where the most power is. Over time—days-week-months—this will reduce the reactivity of the centre of the core compared to the edges, and the power shape will change. There will be a slow movement ('redistribution') of the power from the centre of the core to the edges, increasing neutron leak-

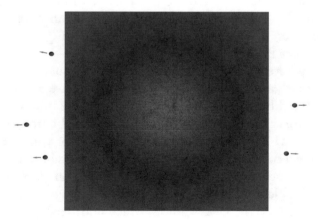

Fig. 8.2 Power shape after refuelling with neutron leakage

Fig. 8.3 Power shape some months later...

age, as you can see in Fig. 8.3. Your total reactor power hasn't changed, just its distribution within the core.

If you compare Fig. 8.3 with Fig. 8.2, you'll see that the leakage of neutrons has significantly increased. Your flux instruments will now see more neutrons and if you don't find a way of recalibrating the instruments you'll be over-estimating your reactor power. You have limits on reactor power (in a rule book), so practically this will prompt you to slowly wind-down the power output. That'll mean your power station is producing less electricity and for no good reason!

That's only one of the problems. Now let's talk about Plutonium.

Plutonium-239 (Pu-239) is a better fuel than U-235. It produces more energy from each fission and tends to provide a few more neutrons. 'But...', you'll be saying '... we haven't put any plutonium in our reactor. We loaded uranium fuel?' Yes, that's true, but there's a process going on in your reactor that manufactures Pu-239.

As you'll have seen in the earlier chapters of this book, most neutrons do not go on to cause another fission—only one does (on average) out of the two or three produced by each U-235 fission. Some of the others leak from the reactor, some are captured by control rods or dissolved boron, and some are captured by the U-238 that makes up around 95% of your fuel. A U-238 atom capturing a neutron becomes U-239. This is unstable and often decays by beta decay (a neutron turning into a proton and an electron) becoming neptunium-239. Neptunium-239 is also unstable and can decay by beta decay to plutonium-239. So the U-238 atoms, which don't readily undergo fission, are actually the source of Pu-239 in a running reactor. This is shown in Fig. 8.4.

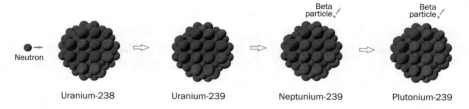

Fig. 8.4 The production of plutonium-239

All U-235/U-238 fuelled reactors make plutonium, and of course, that's the purpose for which they were initially designed as Pu-239 can be used to make nuclear weapons without enrichment technology. Plutonium can be chemically separated from uranium if you have a huge and very well-shielded chemical plant to do it in.

In a PWR, fresh fuel usually contains no plutonium (though it is used in some plants mixed with uranium as Mixed Oxide Fuel). On the other hand, irradiated fuel will contain some Pu-239, and some of the reactor power will be coming from this plutonium. It is both produced in the reactor (from U-238) and fissioned by slow neutrons, just like U-235. There will be a balance between these two processes but, on average, the amount of Pu-239 in the core will rise with fuel burn-up—the longer you run the reactor between refuellings, the more Pu-239 it will contain and the more power you'll be getting from this fuel that you've made yourself!

So why have I talked about Pu-239 production in a chapter on measuring reactor power? Because, on average, fast neutrons produced from Pu-239 fissions will be travelling faster (have higher energy) than neutrons produced from U-235 fissions. The physicists call this effect 'flux hardening', as a faster neutron will hit you 'harder'. Faster neutrons will also travel further in water before slowing down, so are more likely to leak from the reactor. You can probably see the problem now? The more plutonium you make (and fission) in the reactor, the higher the neutron leakage. So just like the redistribution of reactor power, flux hardening will fool your instruments into thinking that your reactor is running at a higher power than it really is.

Incidentally, if you are intent on making nuclear weapons from plutonium, you're only going to be running your PWR for short periods between refuelling shutdowns. Pu-239 sometimes captures a neutron without causing a fission, turning it into Pu-240. This process can carry on into Pu-241, Pu-242 etc. The higher the burn-up of the fuel, the more of these heavier plutonium nuclei will be present. I've been told that they make nuclear weapons behave in unpredictable ways (!), which is why high burn-up irradiated fuel from commercial PWRs is rarely involved in weapons programmes.

There's more, but this one takes a bit of thinking about.

When you start-up your PWR for the first time, your fuel is a mixture of a few percent U-235 and lots of U-238. As it runs, it produces Pu-239, but not enough to offset the number of U-235 atoms that are being fissioned. The reactor will be producing radioactive fission products, some of which will capture neutrons. Overall the reactor will be becoming less reactive the longer it runs. This is a significant effect; a drop in reactivity of around 20,000 milliNiles (20 Niles) between refuellings—see Chap. 3 if you want a reminder of what a milliNile is.

Now, imagine you're a neutron recently produced by a fission and successfully slowed down in the moderator. You find your way back into the fuel. If it's fresh fuel, then around 1 in 20 of the uranium atoms you might meet are going to be U-235. If it's irradiated fuel, that number will be a good deal lower (even allowing for the build-up of Pu-239), and you'll have neutron capturing fission products to contend with as well. Overall, there is less chance that you will cause another fission.

The physicists call this process 'a fall in the macroscopic fission cross-section' if you want to research it further. The upshot is that to get the same number of fissions per second (the same power level), you'll need more neutrons flying around (a higher neutron flux). In your PWR, you can achieve this by slowly reducing the concentration of dissolved boron in the primary circuit, thereby allowing more neutrons to be successfully moderated. In reactivity terms, we'll be reducing the negative reactivity provided by the boron. This means that we can keep the reactor critical at the same power level, even though we are burning-up the fuel.

Unfortunately, with more neutrons moving around the reactor, more neutrons will leak out of the edges. Once again, we've hit a problem that will cause your flux instruments to slowly drift upwards.

It's worth stopping and taking stock at this point. We've seen that measuring flux, that is neutrons leaking out of the reactor, is a good measure of the reactor power in the short term but that it suffers from three longer-term problems:

- power redistribution due to burn-up
- flux hardening from Pu-239
- flux rise due to a fall in macroscopic fission cross-section

Perhaps we need to find another way of measuring power?

8.2   Nitrogen-16

Water-cooled reactors such as PWRs have an alternative power measuring technique—they can use nitrogen-16.

A bit of weird physics here: Water is H_2O, so with 20 tonnes of water flowing through the reactor every second, we can be confident that there are a lot of oxygen atoms moving through the core. Oxygen is mostly oxygen-16 (O-16), made up of 8 protons and 8 neutrons. It turns out that if an O-16 nucleus is hit by a fast neutron, there is a small chance that it will capture the neutron and throw out a proton—I said it was weird. That turns the oxygen-16 into nitrogen-16, with 7 protons and 9 neutrons. Nitrogen-16 is pretty unstable and will decay by beta decay back to O-16 with a half-life of just 7 s. As it does so, it emits a gamma ray at a specific energy. This process is shown in Fig. 8.5.

At this point, it's worth noting that N-16 gammas are one of the big hazards from a running water-cooled reactor such as a PWR or BWR. The gammas are emitted from the water wherever it is in the primary circuit (which includes the steam system on a BWR). If the reactor is running, you can't stand next to the loop pipework, even if the reactor itself is well shielded. On the other hand, the short half-life means that it's all gone in just a few minutes after shutting down the reactor.

N-16 is useful. Your PWR (like many others) is fitted with gamma detectors on the hot legs which are tuned to detect N-16 gamma rays. The more power we have in the reactor, the more N-16 we'll be producing because there will be more fast neutrons. It doesn't matter where in the reactor that power is produced as all of the water goes along the hot legs regardless of whether it travelled through the edge or the middle of the reactor.

But (there's always a 'but'), N-16 production is affected by both 'flux hardening' and the fall in 'macroscopic fission cross-section'. As the core is burnt-up, the number of neutrons has to increase to sustain the same power, so there is a higher chance that an O-16 atom will be a hit by a neutron. Similarly, it

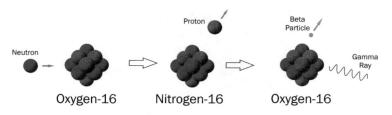

Fig. 8.5 Nitrogen-16 production and decay

turns out that the O-16 to N-16 reaction is more likely to happen with faster neutrons and we'll get faster neutrons as we increase the amount of Pu-239 in the core. So, although N-16 gamma is unaffected by the power shape, it will tend to drift upwards with burnup, just like the flux instruments.

8.3   Using Heat (Primary Circuit)

The nuclear physics appears to have let us down, so let's try something a bit more basic. Stop and look at the PWR in Fig. 8.6.

We've got colder water going in (Tcold) and hotter water coming out (Thot). We can measure the flow of water flowing through the loops, so we know the total flow. We can look-up how much heat is needed to raise the temperature of the water by one degree (the 'Specific Heat'), so the total heat produced by the reactor (its power level) could simply be calculated from:

$$Power = (Thot - Tcold) \times Flow\ Rate \times Specific\ Heat$$

This is called a 'Primary Calorimetric'—Primary because it's a calculation of primary circuit conditions and calorimetric meaning a 'heat-measurement'. It's a quick, simple calculation. But it doesn't work.

Well, that's not entirely true. It does work; it's just not very accurate. Look at the individual terms in the calculation. The specific heat has a good accuracy as many thousands of experiments have gone into measuring the

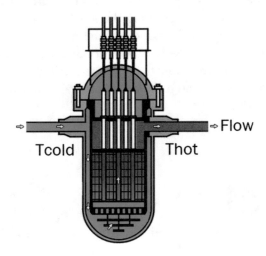

Fig. 8.6   Inputs to primary calorimetric calculation

properties of water. The flow rate is measurable to a reasonable accuracy using simple instruments attached to the loops, which can be calibrated using fancy techniques such as ultrasonics.

You'd think that the measurement of Thot and Tcold would be the best of all, as there are lots of instruments that could be used to get an accurate temperature measurement. In the cold leg, there's a pump spinning very fast, thoroughly mixing all of the water that passes through it. You put a temperature measuring device anywhere in the cold leg, and you'll get a temperature measurement that is representative of all of the water flowing in the leg.

Here's the problem: that's simply not true for the hot leg. The water comes out of the core at a range of temperatures depending on the power of the fuel assembly that it flowed through. There's minimal mixing of the water above the core, as it heads out to the hot legs. Once in the hot leg, there'll be a tendency for the water to 'stratify', with the warmer water at the top of the leg, and the colder water at the bottom, but this won't be stable. The layers will swirl, twist and move around. The result is that you can't be confident that any particular Thot measurement is genuinely representative of the water in the hot leg.

This matters because there's only a 30 °C difference between the hot and cold legs. If you get the measurement wrong by even 1 °C, you will be introducing an error of 3% power. The primary calorimetric is a rough indicator of reactor power, but you wouldn't want to use it to calibrate your flux and N-16 instruments.

8.4 Using Heat (Secondary Circuit)

There is another way: the 'Secondary Calorimetric' calculation. In other words, a heat calculation based on what is happening in the secondary circuit, specifically the steam generators. In steady operation, all of the heat produced in the core will end up in the steam generators, so in principle, we could use what's happening in the steam generators to indirectly measure reactor power. A simple diagram of a steam generator is given here in Fig. 8.7.

Let's assume that you can get a reasonable measurement of feedwater temperature and flow—the pipes are smaller than the hot legs, and the water is well-mixed by the feed pumps, so this is reasonable. Now let's measure the steam pressure; again, easy to get a representative measurement as steam really doesn't stratify. We don't need to know the steam temperature (Tsteam)—that's fixed by the pressure (as you'll see in Chap. 10), nor the flowrate if we assume that all of the water that goes into the steam generator turns into steam.

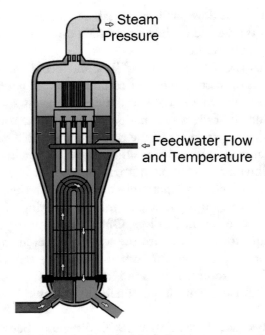

Fig. 8.7 Inputs to secondary calorimetric calculation

To do the actual calculation, we have to use a property of water/steam called 'Enthalpy'. Enthalpy is a word physicists and chemists use to describe the total heat contained in something. It's a bit more complicated than just using temperature, but again we can look this up in tables for our water and steam as it's been measured thousands of times. The secondary calorimetric calculation for an individual steam generator then becomes:

$$Steam\ Generator\ Power = (Enthalpy\ of\ Steam - Enthalpy\ of\ Feedwater) \\ \times Feedwater\ Flowrate$$

We can add up the powers from the four steam generators and make a few easy corrections—such as subtracting the heat added to the primary circuit by the reactor coolant pumps. Then, provided the reactor is running at steady power, the power taken from the steam generators matches that produced by the reactor. Perhaps surprisingly, this secondary calorimetric calculation, performed continuously by a computer using data averaged over a few minutes, turns-out to be the most accurate way of measuring reactor power, with an overall accuracy of around ±1%. You can now drive your reactor using your secondary calorimetric calculator, and you can drive it right up against the power limit in your rule book (which will have taken account of the accuracy of the calculation).

The secondary calorimetric is going to be too slow to use for protection purposes, but we've got the flux and N-16 instruments we can use for that. We can keep those working correctly if we use the results of the secondary calorimetric to check and adjust their calibration every day or so. This means that the flux and N-16 instruments can be used to tell the protection system to automatically shut down the reactor within just a few seconds of a fault, and you can be confident that they won't do this too early or too late.

8.5 What Doesn't Work

As a reactor operator, if you're talking to someone about measuring reactor power, you'll probably get the question: "Why don't you just use the electrical output from the turbine generator to tell you the power?" The answer to this one is that the output of the power station depends very much on what is going on in the secondary circuit. You only need one valve to be passing (leaking), or a feed heater level controller to be playing-up and you'll find that the electrical output will shift by a few megawatts. For this reason, it's pretty unreliable as a measure of reactor power. On the other hand, if your secondary calorimetric is working well, you might spot small changes in overall efficiency that'll point you towards just this sort of problem with the turbine.

8.6 Back to Fissions

Now that we're confident about measuring your reactor power, we can answer the question: How many U-235 fissions does it take to run the reactor at full power? I gave you a number at the start of this book, now I can show you where it comes from.

According to the nuclear physicists, each fission of U-235 releases around 200 MeV of energy. This number includes the energy that you're getting from the decay of radioactive fission products, which is around 6.5% of the total heat from your reactor. If you've not studied physics or chemistry, you might not have seen the MeV before. It stands for Mega-electron-volt, an electron-volt being a standard unit used to measure energy on an atomic scale. 200 MeV is 32 millionths of a millionth of a joule (or 3.2×10^{-11} J), which doesn't sound a lot but is staggering compared to the sorts of energy you get from chemical reactions.

Your PWR runs at a full power of 3500 MW, so all we have to do is to divide the total power by the energy per fission, and we get:

$$\frac{3,500,000,000 \text{ W}}{3.2 \times 10^{11} \text{ J}} = 1.1 \times 10^{20} \left(\text{fissions}\right) \text{ per second}$$

That's just over 100 million million million fissions per second. Maybe your PWR has 50,000 fuel pins in a core made of around 200 fuel elements. Each fuel pin could contain 400 fuel pellets. That's 20 million fuel pellets in your reactor. At full power (on average) every pellet will be responsible for five trillion fissions per second. That's where the real power is coming from; however you choose to measure it.

9

Your Reactor Is Stable (Part One)

This chapter is about 'Reactor Stability'. This is the second of the three key concepts that I mentioned at the beginning of the book.

Let me put it simply: Your PWR Reactor is stable.

What do I mean by that? I mean that, for example, if the reactor power goes up a little bit, the reactor temperature will go up. If the temperature goes up, the effects on reactivity will tend to drive the reactor power back down to where it started. The same happens in reverse if the reactor power falls. You can imagine it like driving a car with springs attached to the steering wheel. The car will keep trying to travel in a straight line, even if you nudge the wheel a little.

For most of the time, this stability makes driving a reactor much easier. It will tend to keep your reactor running at a constant power unless you deliberately change its running conditions. Just occasionally, you'll need to fight against this inherent stability to get your reactor to do what you want. In our car analogy, what would you do if you wanted to turn a corner?

In this chapter, I'm going to explain the two bits of physics that keep your reactor stable in its day-to-day operation and beyond. I'll also reveal a third reason for your reactor's stability that is present in a PWR but is absent in a few other reactor designs (and that contributed to the accident at Chernobyl in 1986).

© Springer Nature Switzerland AG 2019
C. Tucker, *How to Drive a Nuclear Reactor*, Springer Praxis Books,
https://doi.org/10.1007/978-3-030-33876-3_9

9.1   Fuel Temperature

When we were discussing the fission chain reaction, you might remember that many neutrons are lost because they are captured by U-238 atoms. U-238 makes up 95% of the fuel, so it might be surprising that a chain reaction can happen at all. The explanation for this is that U-238 nuclei only have a good chance of capturing neutrons if they arrive at particular speeds, or energies. This is illustrated in Fig. 9.1—you can see how as neutron energy falls from right to left (the neutrons are 'moderated') there is an overall rise in the likelihood that they will be captured by U-238. However, there's a much more dramatic effect at specific energies where capture is very likely indeed. We call these 'resonance capture peaks', and they can steal neutrons from the chain reaction if the neutrons re-enter the fuel during the moderation process.

The peaks in Fig. 9.1 look dramatic, but they're even more impressive when you realise that each mark on the axes of this graph represents an increase of a factor of ten in the likelihood of capture (it's what the mathematicians would call a 'log-log' graph).

Now for a bit more obscure physics. Atoms don't stay still. They vibrate and the hotter they are, the more they vibrate. If an atom is vibrating, then its nucleus must be moving. So, if a neutron is moving towards it at a speed

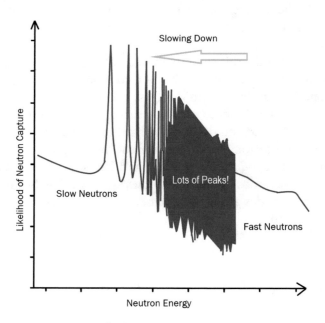

Fig. 9.1 Resonance capture peaks

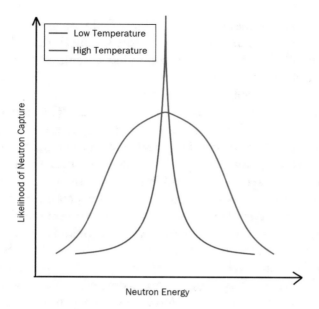

Fig. 9.2   Resonance capture peak broadening

(energy) just off to the side of one of these resonance capture peaks, then there is a chance that the nucleus's movement and the neutron's incoming speed, added together, might just hit the peak. This would dramatically increase the likelihood that the neutron will be captured.

With lots of U-238 atoms vibrating and lots of neutrons going into the fuel with a whole range of speeds, the resonance capture peaks shown in Fig. 9.1 will each appear to get wider as fuel pellet temperature increases. In other words, the hotter the fuel pellets, the more neutrons will be lost. This effect (for a single resonance capture peak) is shown in Fig. 9.2. Strictly speaking, the peaks also get shorter, but that is not a noticeable effect when there are so many U-238 atoms in the fuel.

You might sometimes hear this called the 'Doppler Broadening' effect. This is because it is caused by the 'relative speeds' of the U-238 nucleus and the incoming neutron—in much the same way that the Doppler effect changes the sound from a moving vehicle; higher-pitched coming towards you, lower as it moves away.

If you think about the effect this has on the fission chain reaction, you'll see that it reduces reactivity with rising temperature. The hotter the fuel, the more negative the effect. For typical conditions in a PWR, this effect is worth around minus 4 milliNiles for every 1 °C that fuel temperature increases. The reactor physicists call it the 'Fuel Temperature Coefficient', or FTC.

Minus 4 milliNiles per °C doesn't sound like a significant effect, but that's because we really haven't thought about fuel temperatures yet. You'll remember from Chap. 6 that the water flowing through the reactor rises in temperature from about 290 °C at the bottom, to about 325 °C at the top. So what temperature is the fuel? Well, it's immersed in the water, so the fuel cladding can't be very much hotter than the water. However, the heat isn't being produced in the cladding. It's being produced by fission in the fuel pellets. These must be running hotter than the cladding to drive the heat out (giving what the physicists call a 'temperature gradient', down which the heat can flow).

The fuel pellets are made from uranium oxide—a ceramic material which is very poor at conducting heat. This means that there has to be a very high temperature in the centre of each fuel pellet to drive out the heat—in other words, there'll be a steep temperature gradient across each pellet. At full power, you'll probably have pellets in the reactor whose centres are running at more than 1200 °C. Even if you were to look at an average of the fuel pellet temperature you'd probably still have a value above 600 °C. This is shown in Fig. 9.3, where the red line indicates the temperature measured across a fuel rod at the position shown by the yellow dashed line.

So, if the reactor power rises from just critical to 100% power, the average temperature in the fuel pellets might increase by 300 °C, even though Thot will only have risen by around 30 °C. The overall negative effect on reactivity

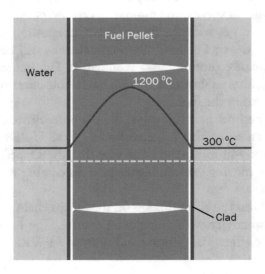

Fig. 9.3 Temperature gradients across a PWR fuel rod

from the FTC will be significant, and you'll have to compensate for this as you raise power, either by withdrawing control rods or reducing the concentration of boron in the primary circuit.

This sounds a bit of a nuisance, but it's great for safety. If anything unpleasant happens and fuel temperatures start to rise, the core's overall reactivity will go strongly negative, and reactor power will rapidly fall. Isn't this just what you'd want to happen?

9.2   Moderator Temperature

The water flowing through your reactor serves two functions. It removes the heat from the core—so you can do something useful with it—and it acts as the 'Moderator' in which fast neutrons can bounce around, slow down and become thermal neutrons. Without the water, you wouldn't have a running reactor.

In common with most chemicals, water expands as it gets hotter. The physics is actually a little more complicated because water goes through its highest-density at just 4 °C, and solid water (ice) is less dense than the liquid, but none of that matters to us at 300 °C. So, lets' just assume that the water in your reactor gets less dense as it gets hotter. What does this do the reactivity?

Here's where there's a clever bit of design in most PWRs.

Your reactor is designed with the fuel rods too close together. Yes, really. If you wanted a better reactor that used neutrons more efficiently and had higher reactivity, you could have one simply by moving the fuel rods further apart.

With the fuel rods closer together than in an ideal reactor, the neutrons are, on average, less well slowed-down than they should be. The physicists would describe your reactor as being 'under-moderated'. If you think back to Chap. 6, you might remember that I described the control rod material (Silver-Indium-Cadmium) as being able to capture neutrons over a wide range of speeds (energies). Now you can see why that's important. One of the consequences of an under-moderated reactor is that you'll have more neutrons moving around that aren't fully slowed-down.

There's a good reason to deliberately design a PWR to be under-moderated; that's stability. When the reactor is running, imagine what happens if the power rises. More heat will be transferred to the water, the water will get hotter, and it will expand. If the water expands, there will be fewer water molecules between the fuel pins, so the water will be less effective at slowing down the neutrons. Starting from an under-moderated position, a heat-up will always give an even less well-moderated reactor. The upshot is that rising

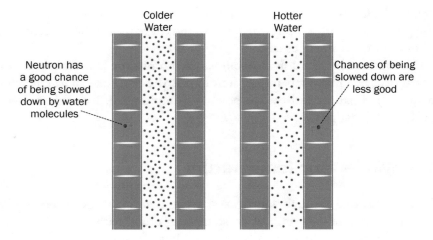

Fig. 9.4 Moderator temperature coefficient

water temperature will reduce reactivity, as you can see in Fig. 9.4 (note that the distance between the fuel pins has been exaggerated in this diagram).

This is a strong effect. In pure water, the Moderator Temperature Coefficient (MTC) can be as substantial as minus 60 milliNiles per °C. If the reactor were designed closer to an 'ideal' reactor, with the rods further apart, you couldn't guarantee this effect in all circumstances.

I'll admit that there is a complication with the moderator temperature coefficient. I gave you an MTC value for 'pure water', but the reactor coolant water has boron dissolved in it. Boron is added to the water to capture neutrons and is used to control the reactivity, a bit like a control rod. In Fig. 9.4, you can see that when the water gets hotter, it gets less dense. But as well as there being fewer water molecules between the fuel pins, there will also be fewer boron atoms. By removing boron you are removing something that captures neutrons. This has an effect on reactivity that is opposite to that of removing moderating water molecules, so it makes the MTC less negative.

At the start of an operating cycle, with lots of fresh fuel in the reactor, there's a need for lots of boron. As the fuel is used up and the core becomes less reactive, the level of dissolved boron is reduced to compensate. This is covered in a bit more detail in a later chapter. What you need to know is with typically high boron at the start of a cycle, the MTC almost disappears; at low boron at the end of a cycle, it's very negative. It changes through an operating cycle, becoming progressively more negative, so you're going to need to know where you are in the cycle to calculate the MTC's effect.

9.3   It's a PWR, so It's Stable

Both the Fuel Temperature Coefficient (FTC) and the Moderator Temperature Coefficient (MTC) have negative values. In other words, they both cause reactivity to fall if the core temperature rises. The FTC is usually the smaller effect of the two, but it is fast-acting and will respond quickly to any rapid change in fuel temperature. The MTC is slower as it takes time for heat to reach the water from the inside of the fuel, but it can be a more significant effect on reactivity as it can be very much more negative than the FTC. (I know that this looks obvious as the MTC is a lot more negative than the FTC but remember that average fuel temperatures move much more than water temperatures.)

If reactivity reduces as temperature rises, then power will fall. This will have the effect of lowering temperature, which will, in turn, bring reactivity back towards zero. For a temperature fall, the reactivity will rise, the power will increase, and temperatures will return to their initial warmer values. This will also return reactivity to zero.

This is a system with 'negative feedback'. It is self-stabilising. Your PWR will be able to run at steady power for days on end with no control rod movement, simply held stable by these temperature effects.

At this point, I should probably point out that there are quite a few reactors built with positive moderator temperature coefficients, including most of the UK's other reactors, the graphite-moderated AGRs (see Chap. 22). This is quite manageable if the moderator takes a long time to heat-up or cool-down, and the coolant has little moderating effect of its own. Reactors built in this way tend to be stable in the short term (because of their negative FTC) and unstable in the longer term (because of their positive MTC), giving time for control systems and operators to intervene.

9.4   Another Coefficient

Reactor physicists like coefficients. There's quite a few more. I'm not going to worry about most of the others here as they don't really affect how you drive your reactor. There's just one that I'm going to explain because it played an important role in the Chernobyl reactor accident.

We know that water becomes less dense as it gets hotter. The most extreme example of this is 'boiling'. If water turns to steam, its density drops by something like a factor of a 1000. Compared with water, steam is a rubbish moderator and is useless as capturing neutrons, so you might expect some significant effects on reactivity if the water in your core starts to boil.

The measure of how much the reactivity changes in a PWR core with boiling is known as the 'Void Coefficient'. Its units are a little different from MTC and FTC, and you'll often see it written down as 'milliNiles per %' where the '%' measures how much of the water has turned to steam (the 'voidage'). In a PWR the void coefficient is strongly negative, usually more than minus 100 milliNiles per % voidage. Let's put it another way: If you get significant boiling in your PWR reactor while it's running, its power is going to drop like a brick. I find that quite reassuring.

Incidentally, this is such a strong effect that it can be quite deliberately used to control the power level in a Boiling Water Reactor (BWR). You'll meet BWRs in Chap. 22. In contrast, AGR reactors use carbon-dioxide gas as a coolant. As the carbon-dioxide is already a gas with very little effect on neutron moderation or capture, AGRs don't have a void coefficient.

9.5   Chernobyl Reactor Number 4, 26th April 1986

Some of the details in the brief description of the accident that follows are 'best-guesses'; based on subsequent modelling and operational experience from other similar reactors (known as RBMKs). There are many books already written about Chernobyl, together with documentaries and dramas—each with a different point of view and sometimes disagreeing on the details. For this reason, I'm not going to try to retell the whole Chernobyl story here, but I am going to point out a link to the physics that you've seen in this chapter. Specifically, the Chernobyl reactor's positive void coefficient was the single most important thing that caused the accident.

You read that correctly. The Chernobyl reactor could have a substantially positive void coefficient if it were operated at low power, with control rods mostly withdrawn.

Why?

As you'll see in Chap. 22, the Chernobyl design of reactor had a boiling water coolant and a graphite moderator, with the fuel and coolant water being separated from the graphite inside pressure tubes. In practice, this meant that it would be possible (in theory) to remove all of the coolant water and still have a viable (if uncooled) reactor. Removing the water removes a significant amount of neutron capture without affecting the graphite moderator.

So… if at low power, an RBMK reactor contains a lot of water that is just below boiling point, a small increase in power and temperature will increase the boiling. This will reduce neutron capture and give the reactor positive reactivity. This will cause power to rise, increasing temperature and giving more boiling. This is a positive feedback mechanism—a positive void coefficient.

On the morning of the accident, a test on Chernobyl reactor number 4 caused just this scenario. Power had been reduced before the test to a point where boiling effectively stopped. The negative reactivity from reducing the steam bubbles (voids) in the coolant effectively 'stalled' the reactor. To compensate for this, and to try to raise reactor power back up to a level at which the planned testing could take place, the operators withdrew far more of the control rods than was usually allowed.

The planned testing involved tripping the turbine associated with the reactor and using its speed while slowing-down to power the pumps circulating coolant through the core. Unfortunately, when this test was initiated at such low power, the reduced flow of cooling water (compared with normal operation) caused an immediate increase in boiling in the core. Reactor power rose very rapidly due to the strongly positive void coefficient. Operators tried to insert control rods to counteract the power rise, but this may have only made things worse as RBMK control rods move quite slowly and had 'graphite followers' hanging beneath them. Inserting control rods from so far withdrawn actually increased the amount of moderator available to the reaction, by pushing the graphite followers into the centre of the core.

The very rapid rise in reactor power caused fuel pins to burst and hot uranium oxide to enter the cooling water. This caused even more boiling. The reactivity increase was so enormous that the reactor easily reached and exceeded 'Prompt-Criticality' (Chap. 4). It's been estimated that the reactor power reached 30,000 MW. That's ten times its designed full power level.

So much steam was produced so quickly that the 1000 tonne lid of the reactor structure was blown off. This 'steam explosion' was followed shortly afterwards by a second explosion, possibly caused by the ignition of hydrogen produced by chemical reactions between the hot fuel and the steam. The exposed hot graphite and fuel then caught fire, spreading large amounts of radioactive fission products out into the environment—a legacy of radioactive contamination that will take many years to fully address.

9.6   Remember That You Have a PWR

This can't happen to your PWR. Your PWR is inherently stable with a strongly negative void coefficient. Any significant boiling in your PWR rapidly reduces reactor power, rather than increasing it.

10

You've Got to Do Something with All that Steam

I'm going to explain the 'Secondary Circuit'—and how you use it to do something useful with the heat that you're taking from the primary circuit. This chapter includes a fair bit of engineering, but you need to understand where all your steam is going. As you'll see later, it can have a surprisingly strong influence on your reactor. Figure 10.1 illustrates the secondary circuit by showing the path from the steam generators, through the 'Turbine' and 'Feedwater Pumps' then back to the steam generators.

10.1 Steam Generators: Viewed from the Other Side...

Let's start by having another look at the steam generators (SGs). When you learnt about the primary circuit, you concentrated on the SG tubes through which the primary circuit water flowed. You'll remember that on the secondary side, water flows up over the tubes, boiling to produce steam. It can do this because it is at a lower pressure than the water in the primary circuit. At full power, the secondary side of the steam generators contains water and steam at around 69 bar and 285 °C (known as 'Tsteam'). This is about 7 °C colder than the lowest water temperature (Tcold) on the primary side, and it's this temperature difference that drives the heat across the tubing. A reminder of the steam generator layout is shown in Fig. 10.2.

The secondary side of the SG contains a mixture of water and steam. At the bottom of the tubing (the 'tubeplate') it's mostly water. By the time it reaches the outside of the top of the tubes it's around 25% steam mixed with 75%

© Springer Nature Switzerland AG 2019
C. Tucker, *How to Drive a Nuclear Reactor*, Springer Praxis Books,
https://doi.org/10.1007/978-3-030-33876-3_10

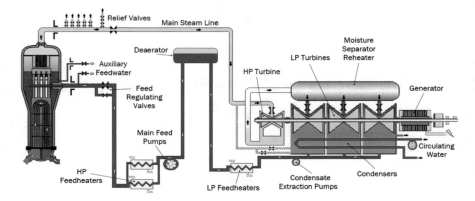

Fig. 10.1 Secondary circuit

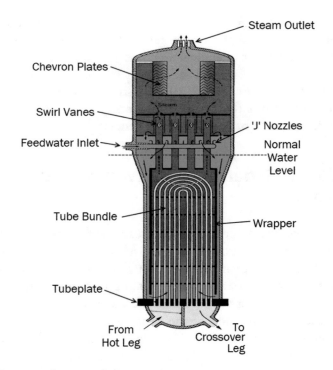

Fig. 10.2 Cutaway diagram of a steam generator

water. This is only possible because the water and steam can exist together on the saturation (boiling) curve that you saw in Chap. 2. We'll come back to this when we look again at reactor stability because it's surprisingly relevant to how your PWR behaves.

A common mistake here is to think that you are pumping water around the outside of your steam generator tubes. You're not. The water flowing in the secondary side of the SGs does so solely through natural convection. It is hotter and less dense within the tube bundle (especially when it contains steam bubbles) so moves upwards relative to the colder water flowing down around the outside. To make this circulation more efficient, the tube bundle is surrounded by a 'Wrapper' that separates the upwards (warmer) and downward (colder) flows.

A 75% water/25% steam mixture isn't what we need for our turbine (we just want the steam), so the next job your SG needs to perform is to separate and dry the steam. In some power stations, this could be done with more heating, but in your SGs the process is mechanical as shown in Fig. 10.2. Firstly the steam/water mixture passes up through 'Swirl Vane Separators'. These look like fixed propeller blades mounted within vertical pipes. As the steam/water passes through the pipes it is made to swirl by the vanes. The water droplets within the steam will naturally spin-out towards the sides of the pipes from where they can drain back down to join the colder water flowing outside the wrapper.

Meanwhile, the now drier steam will carry on up into the 'Chevron Plate driers'. Here the steam is forced to repeatedly change direction by the zig-zag nature of the plates. Any remaining water droplets will tend to be captured at the zig-zags, and drier steam will pass through.

It doesn't sound like it would be a very effective process, but it is. Starting from a mixture that is 75% water, we end-up with steam leaving the chevron plates containing less than 0.5% water. Looked at another way, on average, any 'bit' of water will tend to circulate around the SG four times before emerging as steam.

Taking a lot of steam from your SGs could dry them out pretty quickly, so you need to be topping them-up continuously with feedwater. At low powers this might be cold water straight from storage tanks, and delivered to the SGs by 'Auxiliary Feedwater Pumps'. In contrast, at full power the feedwater will be preheated and will be supplied by the 'Main Feedwater Pumps' (more about that later). How much feedwater? Well, at full power each SG will be boiling around half a tonne of water into steam every second. So your feedwater system has to be able to top-up the four SGs with a total of about 2 tonnes per second, just to keep water levels stable.

Feedwater is pumped in above the wrapper through 'J' shaped nozzles ('J Nozzles') so that it is evenly distributed—but remember this feedwater flow is small compared with the natural circulation flow that is already taking place within the SG. Perhaps it's also worth noting that it's not easy to measure a water level in the steam and water mixture surrounding the SG tubes, so don't try. Instead, when we talk about SG water level we'll measure it in the (colder) water outside and above the wrapper. I've marked a normal water level for the SGs in Fig. 10.2 so that you can see what I mean.

10.2   Main Steam Lines

Now that you understand how steam generators produce steam, you'll want to know where it goes. The steam leaves your SGs through nozzles in their tops then travels along the 'Main Steam Lines'—at full power, the steam will be moving at more than 60 mph. The main steam lines leave the reactor building through carefully engineered penetrations in the building walls. I say carefully because these steam lines are sat at around 285 °C (Tsteam) and you couldn't let them come into direct contact with the walls of the building without damaging the concrete. An example of a main steam line is shown in Fig. 10.3.

In the early days of steam engines, a few rather nasty events occurred when boilers containing steam reached too high a pressure and exploded. These explosions released a lot of steam, but more importantly, they tended to throw bits of boiler very far and very fast; often resulting in fatalities. Such events led to the universal adoption of 'Safety Valves' which open automatically to release steam if the boiler pressure gets too high. Your PWR is protected in the same way. On the primary circuit, safety relief valves of several different designs can be attached to the nozzles on the top of the pressuriser. On the

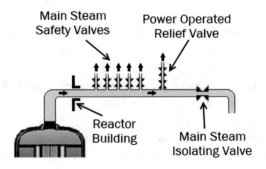

Fig. 10.3 Main steam line

secondary circuit, the safety relief valves (known as 'Main Steam Safety Valves', MSSVs) are attached to each of the main steam lines, outside of the reactor building. As there are no valves that can be closed in the main steam lines between the steam generators and these MSSVs, they will always be able to prevent the secondary side of the steam generators from being over-pressurised. Being outside of the reactor building, the MSSVs can discharge steam through vent pipes, directly to the atmosphere.

Your PWR has two different kinds of relief valves on the main steam lines. There are around half a dozen MSSVs on each main steam line, with a single 'Power Operated Relief Valve (PORV)'. Each of the MSSVs is held closed by a large spring and will only open if the steam pressure is high enough to overcome the spring force. These valves probably weigh 1/3 of a tonne each, so the springs really are 'large'. Typically, the MSSVs won't lift unless steam pressures rise above 80 bar, though each valve spring is set a little differently from its neighbours. Why this complexity? Well, it ensures that the minimum number of MSSVs lift in any over-pressure event, reducing the risk that one might then stick open. Also, by having multiple smaller valves rather than one large valve, the cooldown effect of a stuck open valve would be reduced.

The PORV is a bit different from the MSSVs. There's just one per main steam line, but rather than a spring, the valve can be opened by the operator at a pressure that can be varied. It is usually set to lift at a lower pressure than the set pressure of the first MSSV. You'll see why the PORV is essential when we talk about steam dumping and cooling-down the primary circuit, later on.

The final thing to see in this bit of the main steam line is the main steam isolating valve (MSIV). If you think about it, you'll realise that closing all four of the MSIVs will isolate each of the steam generators from the others. This is why each main steam line has to have its own set of MSSVs and a PORV. If you like, the MSIV marks the end of the 'nuclear' bit of each line. After this point, the steam lines are really just the same as the conventional pipework that you'd find in a non-nuclear power station.

10.3   Steam Turbines

This isn't a book about steam turbines. There's lots of material out there if you want to research the subject in more detail (for example, look-up the steam-ship 'Turbinia'; you'll be glad you did). Instead, I'm going to give you an overview of how your turbine will be working for you.

Your PWR has a single large Turbine-Generator set. The turbine is the bit that the steam drives; the attached generator produces the electricity. Smaller turbines (say, those producing less than 700 MW of electricity) will be synchronised with the grid and spin at the same speed. In other words, if, as in the UK, your grid uses alternating current at 50 Hz (50 cycles per second), then the turbine generator will be spinning 50 times per second, or 3000 revolutions per minute (rpm). That's pretty impressive when the spinning bits weigh more than 300 tonnes and has blades (see below) travelling at twice the speed of sound. It's even more impressive when you discover that they can do this for a couple of years at a time, with no shutdowns in between. Steam turbines have been around for more than a 130 years and are very reliable machines, if they are looked after.

There are practical limits to how big turbines can be made. A 1200 MW turbine spinning at 3000 rpm would be very expensive to build. The forces acting on the turbine blades would be much higher than in a 600–700 MW machine so the engineering would have to be considerably more robust. Instead, large turbines—including yours—typically spin at half the frequency of the grid—in the UK that would be 25 Hz or 1500 rpm. The generator is wound differently, with twice as many magnetic 'poles' compared to a 3000 rpm machine, so the result is that the generator still produces electricity at 50 Hz.

Usually, when you hear about a turbine you are told that it's like rows of propeller blades on a shaft. If you let steam in at one end, it's going to push on each row of propellers in turn, so spinning the shaft. The inventors of early steam turbines realised that things weren't quite this simple… The first row of blades causes the steam to spin along with the blades; so the second row of blades gets hardly any push. To get useful power from all the rows of blades, you need to straighten out the steam flow in between the rows of moving blades. This is done by introducing 'fixed' blades that are attached to the casing and don't spin, in between the rows of moving blades attached to the shaft.

Next, you have to remember that as steam gives up its energy to the spinning shaft, it's going to lose temperature and pressure. As its pressure drops, it will expand. This means that your turbine casing will need to be shaped like a cone, with blades getting bigger in each subsequent row, to be able to make use of the expanded steam.

Figure 10.4 shows a cutaway of a steam turbine that has been built with the above ideas in mind. You can see the shaft, the moving blades and the fixed blades—in this case, two sets of each.

You've probably realised that there will need to be some kind of sealing arrangement around the shaft; otherwise the steam would escape without pushing on the blades. In Fig. 10.4, in common with most large turbines this sealing is provided by a 'gland' into which steam is supplied—some flows in, some out, sealing the turbine in the process (a bit like the seal injection flow supplied to your reactor coolant pump shafts).

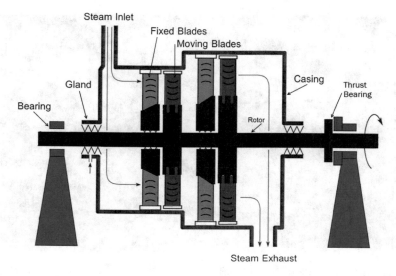

Fig. 10.4 Simple steam turbine

Finally, it's worth noting that as the steam pushes on the blades to make the shaft spin, it also pushes along the shaft. You could build a turbine like the one shown in Fig. 10.4, but you'd need a substantial bearing behind the blades to absorb this lengthways thrust. Instead, a more typical design for powerful steam turbines is to have two sets of blades mounted opposite to each other. Unlike the turbine shown in Fig. 10.4, the steam would then enter at the middle and expand (push) in both directions, so balancing out the thrust. This is how your real 'High Pressure (HP) Turbine' is constructed.

10.4   The High Pressure Turbine

Steam travels along the main steam lines and then through a set of valves (known as 'Stop' and 'Governor' valves) after which it can pass into the blades of the HP turbine. The governor valves are designed to throttle the amount of steam going through; the stop valves are fast-closing valves that are only used when you want to isolate the turbine from the incoming steam, such as when you're shutting the turbine down.

Figure 10.5 shows an example of a set of moving HP blades being lifted from a typical (600 MW) turbine during overhaul. You can get an idea of its size by looking at the handrails on the left of the picture. Your HP turbine is bigger than this but has a very similar shape.

Remember, the steam enters this turbine at the middle and expands (pushes) in both directions, causing the shaft to spin, so balancing out the thrust.

Fig. 10.5 High pressure turbine blades

10.5   Re-using the Steam

After passing through the HP turbine, your steam will be colder, lower in pressure and much wetter. In theory, you could send this steam straight into another (much bigger) steam turbine... but it wouldn't last very long. All of those water droplets hitting blades spinning at 1500 rpm would do a lot of damage. What we really need to is to heat the used steam up a bit, drying it out. Sending the steam back into the steam generators to be re-heated would involve some complicated pipework, so your PWR takes an easier route.

A small amount of the steam from the main steam lines (known as 'Live Steam') is diverted to heating coils that are used to reheat the steam exhausted from the HP turbine. In fact, to make this even more effective, it's a two-stage process with some steam 'bled' from partway along the HP turbine, before being sent to another (cooler) stage of heating coils. Both sets of coils are held in a tank, with HP exhaust flowing over the coils after flowing through a second set of chevron plate driers, as you can see in Fig. 10.6. This is called a 'Moisture Separator Reheater', and it is common-place on PWRs which produce lots of (relatively) low temperature and pressure steam.

Now that your steam has been reheated it passes through a new set of stop and governor valves before entering a much larger 'Low Pressure (LP) Turbine' (Fig. 10.7). The LP turbines are much larger than the HP turbine as the steam has expanded while passing through the HP. In practice, it's usual to have more than one LP turbine, attached to the same shaft as the HP turbine, just

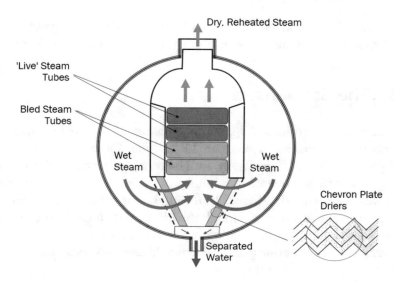

Fig. 10.6   Cross-section through a moisture separator reheater

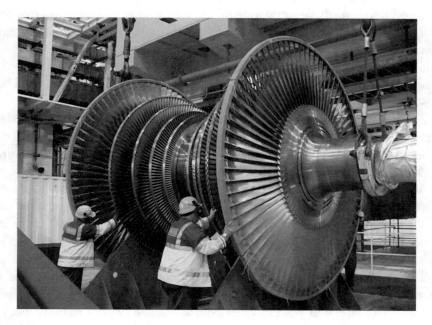

Fig. 10.7   Low pressure turbine blades

to cope with the volume of the steam. Once again, the steam passes through rows of fixed and moving blades, pushing the shaft around as it does so.

10.6   The Condensers

On smaller steam turbines, the steam is sometimes exhausted straight to the atmosphere after passing through an LP turbine. On a larger turbine, this would give us two problems. Firstly, we'd be wasting water, and this water will have been cleaned and treated to minimise corrosion in the secondary circuit, so that would be expensive. Secondly, it would limit the amount by which the steam can expand as it passes through the LP turbine. It couldn't expand to a pressure below 1 bar (atmospheric pressure) as this would prevent it from being exhausted to the atmosphere. This would seriously affect the amount of energy we could take from the steam.

Instead, large turbines such as yours include a 'Condenser'. In engineering terms, this is just a colossal bank of cooling tubes (tens of thousands of them) arranged underneath, or to the side of, the LP turbines. Steam leaves the final row of LP blades and is then directed into the condenser. There, the steam condensers on the cold surface of the tubes. By condensing the steam back into water, you've made it much easier to pump. Like many other PWRs, your condenser is cooled by pumping seawater ('Circulating Water'), at say 10–20 °C, through the tubes.

The clever bit is that if you first attach a vacuum pump (an 'Air Extraction Pump') to the condenser to remove all of the air from the space around the tubes, the pressure in the condenser will fall to the boiling pressure of water at the temperature of seawater. In pressure terms, this is anywhere from 10 to 50 millibars, or less than 1/20th of atmospheric pressure. I say that's the clever bit because it means that the steam can expand all the way to this very low pressure, so you can extract much more useful energy from the steam to turn into electricity. Your condensers really do 'suck' (and that's a good thing).

To give long life and corrosion resistance, modern condenser tubes tend to be made from titanium. They aren't cheap, but they are resilient and will probably still have a high scrap value even at the end of your station's life. Of course, you need lots of seawater to be pumped through the condenser tubes to remove all the heat from the steam—a few million tonnes per day, and that's going to mean large circulating water pumps.

10.7 The Way Back

To get the water (condensed steam) back to the steam generators requires two things: pumps and heaters. Often there's more than one stage of pumping, but however your feedwater pumps are arranged, you'll need to achieve a flow of around 2 tonnes of water per second, travelling back to the SGs at SG pressure (70–80 bar). To give you an idea of scale, your PWR might be using 25 MW of electricity just to run the feedpumps.

You'll see in Fig. 10.1 that there are several sets of 'Feedheaters' in the path back to the SGs. Heating the feedwater reduces the stresses on the metal components inside of the SGs as you won't be continually cycling their temperatures. More importantly for you, it makes a huge difference to the overall efficiency of the power station. Perhaps surprisingly, heating the feedwater before it is sent back to the steam generators actually improves how heat is used in your power station (the 'thermal efficiency'). The reasons why are a bit beyond this book, but if you're interested, do some research into 'Steam Cycles'.

Your feedheaters use steam bled from the LP and HP turbines in the same way as the moisture separator reheater used steam bled from the HP. The 'drains' (condensed steam) from the bled-steam side of each heater are cunningly connected to one another to recover as much useful heat as possible (e.g. drains from hot heaters assist in heating earlier (colder) heaters).

The 'Deaerator' (shown in Fig. 10.1) is a special feedheater whose other role is to remove dissolved gases from the feedwater. Height is important here. The deaerator is mounted high up in the turbine hall to provide a raised pressure at the inlet to the 'Main Feed Pumps'. That means that water has to be pushed up to the deaerator by the 'Condensate Extraction Pumps', which are all the way down in the basement, at the bottom of the condensers.

Compared with, say, a coal-fired power station, your PWR runs at relatively low temperatures. Physics says that the maximum theoretical efficiency of a process using heat is related to how far apart its highest and lowest temperatures are. For a coal station, this temperature difference might be 600 °C. For a PWR it's nearer 300 °C. The maximum theoretical thermal efficiency for a PWR would be just 50%, i.e. you could turn 50% of the energy you produce in the reactor into useful work. In reality, there are losses in the process, and the best you'll achieve is around 35% or a little less. Put another way, roughly two-thirds of your reactor heat is dumped into the seawater when you are running your PWR at full power. There's nothing you can do about it… you can't change the laws of physics (Captain?)!

Finally, before reaching the SGs, the feedwater passes through 'Feed Regulating Valves' which are used to control the flow of water. These are discussed in more detail in Chap. 16.

10.8   The Generator

You'll have noticed that so far all we've achieved is to turn the turbine shaft—admittedly very quickly. At the other end of the shaft from the HP turbine is a large electrical generator. At your station, the outside of the generator (called the 'Stator' because it doesn't spin) is wired through a big switch (a circuit breaker) and then via transformers to the electrical grid. As the grid uses alternating current (AC), this has the effect of creating a magnetic field within the stator that spins around synchronised with the grid.

The turbine-generator shaft, spinning because of the steam expanding through the turbines, passes through the stator. Here the shaft is known as the 'Rotor', and it is wired to produce a magnetic field of its own. How do you create a magnetic field in the rotor? By putting a smaller generator, usually called an 'Exciter', on the same shaft. Sometimes the exciter even has its own smaller generator known as the 'Pilot Exciter'.

Fig. 10.8   600 MW turbine generators

To illustrate this arrangement, Fig. 10.8 shows two large (600 MW) turbine generators side-by-side. The small HP turbines would be on the left-hand side of the LP turbines but are hidden by the pipework. Remember, your single turbine is larger than these, but the layout of equipment is very similar.

Your power station puts its energy into the electricity grid by aligning (synchronising) the magnetic field of the rotor with that of the stator and then trying to push against the stator's field to make the alternating current cycle at a higher frequency in the grid (see Chap. 13). This interaction with the grid has important implications for how you run your reactor, as you'll see later.

10.9   The Big View of the Power Station Cooling Circuits

You'll now understand that your PWR Power Station has three main 'circuits', for moving heat.

- The primary circuit pumps push water through the reactor core and into the steam generator tubes before returning it to the reactor. It is a 'closed-loop' system, using the same water over-and-over again, albeit with some adjustments from the CVCS. This keeps potentially radioactively contaminated water separate from the non-nuclear bits of the power station.
- The secondary circuit takes steam from the steam generators, expands it through turbines, condenses it back into water, reheats it and pumps it in to the steam generators where it is boiled all over again. This is also a 'closed-loop' system, keeping water that has been expensively treated and dosed from simply being thrown away.
- The third cooling system, known as the circulating water system, is the 'open' one. It's the system that pumps the seawater through the condenser tubes and back out to sea, perhaps 10 °C warmer than when it came into the station.

Around 1/3 of the energy your reactor produces is turned into electricity sent out onto the grid. The other 2/3 leaves the power station as waste heat in the seawater. Never mind.

11

The Big Red Button…

It won't always be red, nor will it always be very big, but every control room will have one: a 'Reactor Trip' button. Figure 11.1 shows an example—note that a key is needed to reset this one, not to use it!

Remember that the control rods are held out of the reactor by electromagnetic grippers, needing an electric current to hold onto them. The reactor trip button is connected to switches—known as 'breakers'—that interrupt that electric current. If you push the button, the breakers will open. There are usually multiple breakers, arranged so that no single breaker sticking closed could allow the current to keep flowing.

Opening the control rod grippers allows the rods to fall into the reactor under gravity. This sounds crude, but it takes less than 2 s for the control rods to fall from fully withdrawn to fully inserted as they only have a few metres to fall. We have to allow a little time for the breakers and grippers to open but even so, from the time you push the reactor trip button to the time that the control rods are fully inserted is probably only 3 to 4 s.

By the time you look up from the button, you'll see that reactor power has already fallen to around one per cent of what it was before you tripped; and it'll continue falling. Of course, the reactor will still be producing a lot of heat (decay heat), but the fission chain reaction has effectively stopped. In other words, the insertion of the control rods has taken the reactor deeply sub-critical.

© Springer Nature Switzerland AG 2019
C. Tucker, *How to Drive a Nuclear Reactor*, Springer Praxis Books,
https://doi.org/10.1007/978-3-030-33876-3_11

Fig. 11.1 The Reactor Trip Button

11.1 What Next?

Whether you've tripped the reactor by pressing the trip button, or it tripped automatically because the protection systems decided it should be shut down, you're going to have to go through the same 'post-trip actions'. You and the automatic systems will both be focussed on stabilising the plant. Once it's stable, you'll have time to step back and think about what you're going to do next, e.g. sort out whatever problem caused you to press the big red button in the first place! But for now your role is to check that the plant is stabilising and to intervene if you see any problems. You're going to need to learn how to do this quickly, and from memory, before you open any procedures.

The first things to check are:

- Check that all of the control rods are fully inserted.
- Check that the reactor power is very low and falling.

If there's a problem, you'll need to find other ways of reducing reactivity, such as quickly adding boron to the primary circuit.

Next:

- Check that the turbine generator has also 'tripped.'

A turbine trip will have been sent to the turbine control system when you tripped the reactor. A turbine trip causes the stop and governor valves on the turbine to close. These valves are held open by a hydraulic system, acting against mechanical springs. A turbine trip signal works by releasing the hydraulic pressure, so allowing the springs to close the valves very quickly.

Why does tripping the turbine matter? Because, if the turbine isn't tripped, your reactor will be shut down, but your turbine will continue taking steam from the plant. This will cause a rapid cooldown, increasing reactivity, and could even take the reactor back to criticality. Remember, you're trying to stabilise the plant, so if the turbine hasn't successfully tripped you'll be closing alternative steam isolation valves (such as the MSIVs) to stop the cooldown.

Just a few seconds after the turbine trip—time for any steam already in the turbine to flow through to the condenser—the main generator will be automatically disconnected from the grid, and the turbine will begin to slow-down. Again, if there's a problem with this, you'll be choosing alternative electrical switches to open. If you leave the generator connected to the grid it could act as an electric motor and keep the turbine spinning, possibly causing damage by overheating.

Now:

- Check that primary circuit temperature and pressure are both stabilising.

Actually, primary circuit pressure will have dropped significantly when the reactor tripped.

This is only to be expected. When you stopped producing power in the core, Thot fell to pretty close to Tcold (decay heat is only a few per cent of full power, so doesn't make a big difference). In other words, the average temperature in the primary circuit has dropped by around 15 °C. This caused the water in the primary circuit to contract, and so the pressuriser water level suddenly dropped. The pressuriser steam bubble has expanded to fill the gap, and its pressure, along with that of the rest of the primary circuit has fallen. So, what you expect to see after a trip is a primary circuit pressure drop and all of the pressuriser heaters turning-on automatically to restore nor-mal pressure.

Perhaps surprisingly, the steam pressure will have risen after the trip (yes, really). If you think back to what you learnt earlier about heat flow from the primary circuit to the secondary circuit, you'll remember that the temperature difference between the two (around 7 °C at full power) is necessary to drive the heat across the steam generator tubes. After the trip, the heat being pro-duced in the reactor drops away, so this temperature difference can shrink to just a few tenths of a degree. In practice this means that the steam temperature rises, from its full power value of 285 °C, to almost match Tcold. If you're not sure why the steam temperature rises, remember that with the turbine tripped, hardly any steam is now flowing from the tops of the steam generators. It makes sense that the steam pressure would build-up—taking steam generator temperature with it.

Here's the clever bit… the steam dumps (or PORVs, see Chap. 12) will now open taking a small amount of steam away and controlling steam pressure at this higher value. That removes heat from the system and stops both the steam temperature and Tcold from rising any further.

If there's any sign that pressure and temperature are not stabilising, then you'll need to step in and perform some of these operations manually.

And relax:

Once things have settled down, and decay heat is reducing, you'll be changing over to lower-flow auxiliary feedwater pumps. You'll also have to check that you have enough boron in the primary circuit to keep the reactor sub-critical in the longer term. Then it's just a matter of either getting ready to re-start the reactor and turbine, or preparing to cooldown and depressurise the plant. Which you do will be dependent on why it tripped, and that means you're probably going to need support from the rest of the organisation to move forward.

11.2   Trips and Scrams…

Incidentally, in the UK we use the word 'trip' for a rapid shutdown of the reactor or turbine, perhaps because in electrical systems, opening a breaker or switch is also often referred to as 'tripping'. In the USA, especially at Boiling Water Reactors, you might hear the term 'Scram' instead of 'trip'.

There is a story that the word 'Scram' was originally an acronym used at Chicago Pile 1 (the world's first man-made reactor that you met in Chap. 4). One of the physicists, present for the first start-up of CP-1, was positioned with an axe, ready to cut a rope that would release control rods quickly into the core. This was just in case the reactor couldn't be controlled. He was allegedly known as the 'Safety Control Rod Axe-Man' or 'Scram' man, and it's possible that this term then became more widely used to denote any emergency shutdown. It's also possible that this term was back-fitted to the history of CP-1 and the truth is instead that 'Scram' was used in the more traditional context of 'run away!'

In France, a 'trip' is simply known as 'une arête' or 'a stop'. You can't really argue with that one…

11.3   What Makes a Good Control Room?

Control rooms vary enormously in size, age and design. I'm used to panels with controls arranged along 'mimics' of every important system. Figure 11.2 is an example of part of one such mimic; in this case, for a 'High Pressure Safety Injection Pump', something you'll meet in Chap. 18. Even not knowing what the pump does, you should be able to see a clear flow path from the tank (and sump) on the left-hand side of the mimic, through various valves and the pump, to the cold legs on the right.

As an operator, these mimics can make a big difference when you're working under pressure, and help to ensure that you're starting the correct pump or operating the right valve.

Other control room designers choose a different approach. Some prefer banks of similar-looking switches arranged in orderly rows, with no apparent 'mimic' of the associated plant systems. Conversely, at more modern plants there might be no physical mimics on most of the panels, yet there could be hundreds of mimics available on large screens. There is no right answer to control design, but there are lots of opinions... What's important is that you are trained thoroughly on whichever you are using; so are aware of all of its features and potential error-traps.

Let's assume that your control room is a little more traditional, with physical panels and mimics. What sorts of controls and indications might you have? Well, Fig. 11.2 has already shown you that you'll have switches that operate valves. There will probably be a few hundred of these—by no means all of the valves on the power station, but at least the important ones that you might need to operate from the control room within, say, 30 min of a trip.

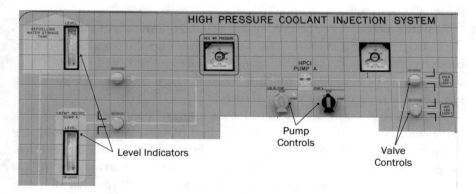

Fig. 11.2   An example Control Room Mimic

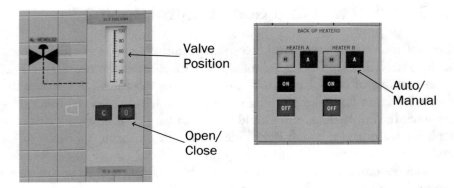

Fig. 11.3 Control buttons

You'll also have starting and stopping controls for pumps; you've got lots of pumps in lots of plant systems. Figure 11.2 reveals an interesting design choice: the pump is shown with green lights. On the plant that this mimic was taken from green means 'stopped' and red means 'running', perhaps on the basis that something that is stopped could be considered to be inherently safe, hence 'green'. Some plants use the opposite idea—red for stopped, green for running. Make sure you know which convention your plant uses!

Sometimes a simple switch for a valve isn't enough as you want to be able to vary its position, rather than just open and close it. You might then have a control plaque such as the one of the left of Fig. 11.3, where the valve movement is controlled by the open and close buttons, but can be stopped at any intermediate position. You might also find buttons to engage or disengage automatic control of equipment. The right-hand side of Fig. 11.3 illustrates this with automatic/manual controls for pressuriser heaters. In this case, you are only able to turn these heaters on or off from the control room when 'Manual' is selected.

As well as controls, your control room will have hundreds of indications on the panels, with tens of thousands more available on the computer systems. Modern nuclear power stations are very heavily instrumented. Panel instruments have to be robust and straightforward to use, with little chance of being misread. A couple of examples are shown in Fig. 11.4.

The indicators on the left of Fig. 11.4 are level and pressure indicators for a pressuriser. Notice how different shaped dials have been used to avoid one being mistaken for the other. The right-hand side of Fig. 11.4 includes a row of green lights. Remember that on these particular panels, green means shutdown, or in the case of a valve, closed. These lights tell you that all five of the safety relief valves on this main steam line are closed.

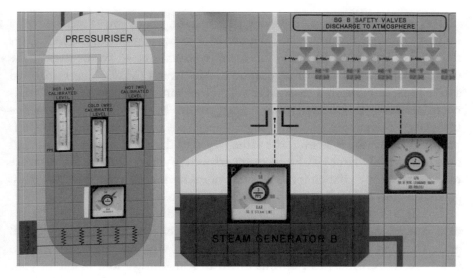

Fig. 11.4   Panel indications

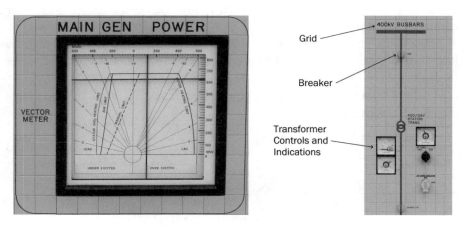

Fig. 11.5   Electrical indications and controls

Importantly, the indications are driven by small switches mounted on the valves—they aren't telling you the position that the valve is supposed to be in, they are telling you its actual position. The event at Three Mile Island (Chap. 19) was partly caused by a control room indication showing the position that the control systems wanted from a valve (closed), rather than the position it was actually stuck in (open). This contributed to the operators in the control room not understanding the real problem that they faced.

This book concentrates on driving your reactor, but as you've seen, there's a lot to do on your turbine as well. Figure 11.5 shows you one of the indicators that is (as far as I know) unique to turbine-generators—a vector meter. The vertical axis shows the turbine output in Megawatts (MW). The horizontal axis is more complicated to explain as it shows something the electrical engineers describe as reactive power, measured in Mega-Volts-Amps-Reactive (MVARs). The people who run the grid need to ensure that reactive power is balanced across the system in the same way that they balance the supply and demand measured in MW. From your perspective you can increase or decrease reactive power by changing the amount of excitation on your turbine's main generator (but only when grid control asks you to!).

The right-hand side of Fig. 11.5 shows part of an electrical mimic—in this case, one of the connections between your power station and the electrical grid. The high voltage grid in the UK operates at 400,000 volts (400 kV), so you have transformers to step this down to something more usable at your power station. In this example, a 'Station Transformer' steps the 400 kV down to 11 kV. You'll also hear people talking about 'Generator Transformers' (step-up) and 'Unit Transformers' (step-down), both usually associated with your turbine-generator.

The switch you can see on the mimic operates a large electrical switch, known as a 'breaker', and could be used to disconnect this station transformer from the grid.

Finally, I'll just mention that, as you can see from Fig. 11.6, turbines have Big Red Buttons too!

Fig. 11.6 Turbine Trip Button

11.4 How Many Reactors?

It's worth taking a step back at this point and asking 'How Many Reactors are there in the world?' If you visit the International Atomic Energy Agency (IAEA) website, you can find a list of 450 operating nuclear power station reactors (including 300 PWRs) in 32 countries, with around 50 more under construction. Many of the existing reactors are getting near the end of their economic lives, so some will probably be shutting down as others are starting up. Worldwide, we've accumulated more than 18,000 reactor-years of operating experience as far as full-scale (power station) reactors are concerned, with nuclear producing a seventh of the world's electricity.

But that's not all of the reactors…

There are around 250 research reactors in 55 countries. These don't generally produce any electricity but are used by companies and universities to research fuels and materials planned for use in larger plants. They are also commonly used to create radioactive nuclei for use in nuclear medicine. Many thousands of people have benefited from these treatments; for example, overactive thyroid disorder is now routinely treated with short-lived radioactive iodine.

Many research reactors are of a simple 'Pool' design, i.e. the fuel (often thin plates of highly enriched uranium or plutonium) sits in the bottom of a deep pool of water. The reactor runs at low power, usually no more than a few kilowatts. The water removes the heat from the fuel and shields the people above from the radioactivity. The 'OPAL' reactor in Australia is an example of a modern pool-type research and medical nuclei production reactor, and (in 2019) is Australia's only operational reactor.

And there's more…

PWRs arose from the idea that a compact nuclear reactor could be used to power submarines. This is still the case, and the concept has been extended to large warships such as aircraft carriers, which might have up to eight individual reactors, though one or two is more common. One estimate of the number of nuclear reactor-powered ships and submarines currently in service worldwide is around 140 vessels, powered by about 180 reactors.

That's nearly 900 operational reactors worldwide. It's not exactly a niche industry, is it?

(I admit, there are some even more obscure reactors that I haven't counted… small, fast reactors used as space power sources; a reactor temporarily installed at an American base on Antarctica; a few possibly fitted to aircraft or missiles etc. I'm not going to worry about them here.)

12

Your Reactor Is Stable (Part Two)

In earlier chapters, I introduced the first two key concepts that help you understand the behaviour of your PWR. The first was the concept of reactivity; the second was the stability of your reactor, especially in response to changing temperatures within the core.

In this chapter, I'm going to explain the third key concept—how the design of the steam generators re-enforces the stability of the plant, and how it dictates what your reactor will do in response to events in the outside world.

I'll admit that this aspect of a PWR's behaviour can sometimes be a bit of a struggle to make sense of. Don't worry; it's easier than it looks. Put simply, a PWR reactor follows the steam demand.

12.1 Steam Generator Conditions

Figure 12.1 is a repeat of the water boiling curve (the 'saturation curve') from Chap. 3. This time I've marked the approximate range of temperature and pressure on the secondary side of the steam generators.

What you'll notice straight-away is that the SG secondary side always sits somewhere on the saturation curve. Having seen the design of the SGs, this should make sense. There is nothing in the design of the SGs that could move their conditions off to the right as that would require something that could superheat the steam. On the other hand, the SGs can't exist to the left of the saturation curve as that would stop them from boiling, and they wouldn't produce any steam.

© Springer Nature Switzerland AG 2019
C. Tucker, *How to Drive a Nuclear Reactor*, Springer Praxis Books,
https://doi.org/10.1007/978-3-030-33876-3_12

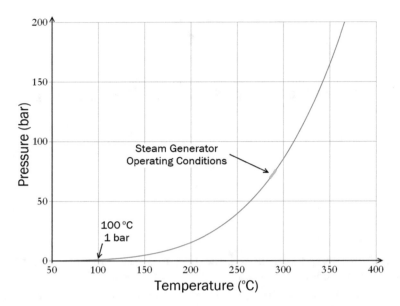

Fig. 12.1 Saturation curve for water

At the heart of the third key concept is simply this: whatever you do to the steam generators, they are only able to move up or down the saturation curve. For example, if you take more steam by opening the turbine governor valves, the steam pressure will fall. This will cause the steam and water temperature within the SGs to fall because they have to stay on the saturation curve. Conversely, if you take less steam, the steam pressure and temperature will rise to stay on the same curve. In all but the most extreme faults, the SGs staying on the saturation curve doesn't change, so it provides a sort of fixed point (or line) around which the behaviour of the secondary circuit is constrained.

This matters because the temperature of the primary circuit water inside the SG tubes (returning to the reactor as Tcold) is closely linked to the temperature outside the tubes (in the secondary circuit). So now we have a link between the steam conditions (Tsteam) and Tcold—remembering that from our second key concept, if Tcold changes, so do reactivity and reactor power.

12.2   Heat Transfer

It's worth stopping and thinking about just how strongly Tcold is linked to the steam generator temperature, Tsteam. There are more than five thousand tubes in each SG. Together they form an immense surface area over which heat can be transferred. You have fast-moving water inside the tubes and at least 75% water (in a mixture with steam) bubbling away on the outside of the tubes—I say at least because there will be less steam at the bottom of the tube bundle than at the top, where it's roughly 25% steam. The tubes are thin—only 1 mm thick—and are made of a metal which is excellent at conducting heat. With such good conditions for heat transfer, it's not surprising that there is only a 7 °C temperature difference between Tcold and Tsteam even at full power. That's with nearly 900 MW of heat passing across a single SG's tubes from the primary to the secondary side.

You might be wondering why I'm talking about Tcold rather than Thot—if Thot is hotter, why isn't Tsteam closer to Thot? The answer is in the inverted 'U' tube design of the SG tubes. In your PWR there is no separation on the secondary side between the hotter side of the tubes, where primary side water enters at Thot and the colder side of the tubes where it leaves at Tcold. This means that Tsteam is going to naturally end-up colder than both Thot and Tcold. If you think about it, you'll see that if this wasn't the case, heat would flow in the wrong direction, from the secondary to the primary sides across the tubes next to where water is leaving the SG.

12.3 A Practical Example: A Small Change in Electrical Power

Imagine your reactor is running near to full power and you decide to raise your electrical generation just a little, say by a few Megawatts (MW). You're going to do this by opening-up the turbine governor valves to admit a bit more steam to the turbine. This will push the blades and the shaft more strongly, which in turn will put more MW of electrical power into the grid.

Now let's think about what is happening to your secondary circuit. As you open-up the governor valves, the path that the steam is taking becomes a little less restricted. This will increase its flow and reduce its pressure—a bit like moving your thumb off the end of a garden hose. This pressure drop propagates backwards along the main steam lines all the way to the SGs whose secondary side pressures will fall. If steam pressure falls, so will Tsteam (to stay on the saturation curve). If Tsteam falls, Tcold will fall. So the first effect you'll see in the primary circuit, from an increase in steam demand—opening the turbine governor valves—is a (small) drop in Tcold.

Incidentally, if I then ask you to tell me what will happen to water levels within the SGs, what would you say? I think that most people realise that they will be taking more water out of the SGs (as steam) than they are putting in (as feedwater) so the SG water levels will fall, won't they? Actually, to begin with, they go up! Remember that you SG water is really a mixture of steam and water? Well, you've just dropped the pressure so all the steam bubbles in the mixture will rapidly expand causing the water level to go up. But only to begin with. After that, the SG water level will start to fall just as you'd expect, so the feedwater control system has to be intelligent enough to recognise what's going on and increase feed to maintain a steady level, despite this seemingly contrary change in level.

What's happening to your reactor? For now, let's assume that the control rods aren't moving so we'll just look at the physics. As Tcold falls the average temperature of the moderator will fall. Average fuel temperatures will fall by a similar amount. From Chap. 9 you'll know that this will cause an increase in reactivity from both the Fuel and Moderator temperature coefficients (FTC and MTC). This will, in turn, cause an increase in power until a new equilibrium is reached, i.e. the increase in power causes a fuel and moderator temperature rise which will pull reactivity back down to zero and so stop the power increase. All of this happens pretty quickly, perhaps a few tens of seconds and it's sensitive enough to respond to changes in Tcold of just a few hundredths of a °C.

What you've just seen is 'the reactor follows steam demand' effect, which is fundamental to the behaviour of a PWR. Once critical, a PWR will change its power level to match whatever steam demand you present it with. There are a few conclusions you can draw from this:

- It doesn't matter to the reactor if the steam demand is coming from steam flow through the turbines, steam flow through a Safety Relief Valve or even steam flow through a broken pipe. If it's there and it affects Tcold, the core will match it.
- Any problem that leads to a drop in Tcold, even if it's only on one SG, will lead to a reactor power rise.
- Any problem that leads to an increase in Tcold will lead to a power drop.
- If you can hold your steam demand steady, your reactor power will be steady.
- If you need to change your steam demand quickly in response to a problem on your turbine or on the electrical grid, your reactor will follow you.

12.4 Keeping on Program

There's a downside to the reactor following the steam demand: changes in temperature. As I've described it, the reason why the reactor will follow steam demand is because of variations in Tcold. Unfortunately, your PWR is designed to operate at a particular temperature. For example, if you increase steam demand and both Tsteam and Tcold fall without being corrected, the effect of the lower temperature steam on the secondary circuit will be to reduce overall thermal efficiency—lower temperature steam isn't as effective in the turbine. Conversely, if you allow Tcold to rise too far as steam demand is reduced, you'll be eating into your margin-to-boiling on the primary circuit.

What you need is a way of tweaking reactivity to bring Tcold back to its programmed (designed) value. You can do this in either of two ways; with control rods or with boron. Moving control rods inwards a few steps or slightly increasing the primary circuit boron concentration (using the CVCS) will cause Tcold to fall to recover 'lost' reactivity. Conversely, a few steps out of your control rods, or a small boron dilution, will allow Tcold to rise to correct the surplus of reactivity. Moving control rods is fast but can have undesirable effects on the power shape in the reactor if you go to far, so there are limits on control rod insertion. Changing the boron concentration in the primary circuit (using the CVCS) gives a more uniform effect but is slower and isn't generally automated.

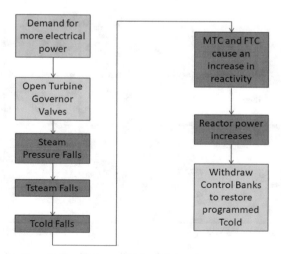

Fig. 12.2 A small increase in electrical power

On your PWR the control rods have an automatic control system known as the 'Reactor Temperature Control System' (RTCS, see Chap. 16) which will step the control rods in and out to keep the reactor at its designed temperature—hence the name! The complete picture for a small increase in electrical power is then given in Fig. 12.2.

From this, I'll leave you to work out for yourself what happens during a power decrease…

Your PWR doesn't actually have just a single value of Tcold. Instead, the programmed value of Tcold is dependent on power. I've already said that there is a 7 °C temperature difference between Tcold and Tsteam at full power. At half-power, with only half as much heat to transfer across the SG tubes, this temperature difference is only 3.5 °C. At zero power Tcold and Tsteam are practically the same temperature. This explains why your steam pressure is highest when your reactor is at the lowest power; that's when Tsteam is closest to Tcold.

It would be better for efficiency to keep steam pressure and temperature higher, but we can't move Tcold too far as we raise power or Thot will end-up too close to boiling. However, we can let Tcold rise a just little bit—say a couple of degrees C—as we raise power. This won't entirely stop the fall in Tsteam, but it will help, as you can see if you compare Fig. 12.3a, b. Your RCTS will have this variation of Tcold built-in as part of its temperature programme, so this should happen automatically as power changes.

In this chapter, I've focused on how steam demand affects the stability of your reactor—the reactor will always match the steam demand from the

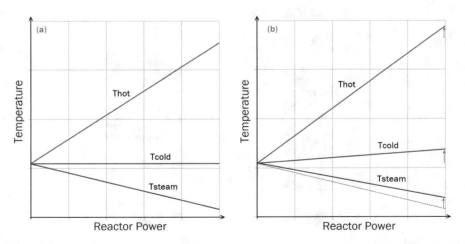

Fig. 12.3 Temperature programme **(a)** constant Tcold programme, **(b)** rising Tcold programme

turbine. But what do you do to control Tcold at low reactor power before the turbine is providing this stabilising steam demand?

12.5   Steam Dumping

The answer is 'Steam Dumping' (Fig. 12.4). First, imagine that your reactor is subcritical with all four reactor coolant pumps in service. Even though the reactor is shut down, it'll still be producing decay heat. If you haven't been shut down for long, this might be more than 10 MW. Added to this, most of the electrical power that goes into running the reactor coolant pumps ends-up as heat in the primary circuit, so in total you might have more than 30 MW to deal with. You'll be using your steam generators to remove this heat as steam, and you won't be able to use it in a turbine (because that won't be up and running yet). So how do you get rid of the steam?

You could simply dump the steam to the atmosphere through the power operated relief valves (PORVs) that we met earlier. In fact, although you have a PORV on each of the main steam lines, you'd only have to operate one PORV to remove all of the heat you're currently producing. They're big valves. You'll remember that the PORVs have a pressure setpoint that you can control from the main control room? So if you drop the setpoint of one of the PORVs to lower than your current steam pressure, it will open and bring the SG pressure down to that value, and hold it there.

The clever bit here is that the PORV setpoint pressure (which you control) corresponds to the pressure that you are going to end up with inside the associated SG (the steam pressure). This also fixes the temperature (Tsteam),

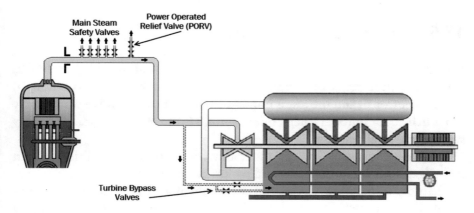

Fig. 12.4   Steam dumping

which in turn fixes Tcold. In other words, by varying the setpoint of the PORV, you are controlling Tcold. The four SGs are connected via the primary circuit water flowing through them and through the core, so if you cool one SG and reduce its Tcold, the others SGs will follow, even if you aren't actively using their PORVs to dump steam.

Using the PORVs in this way is simple but wasteful—your carefully treated water is being dumped to the atmosphere. It's also pretty noisy, so your neighbours won't like it… There is another way: rather than using the PORVs, you can dump steam directly to the condenser of your turbine, without it passing through the turbine blades. This is known as 'Turbine Bypass'. Each part of your turbine's condenser has its own set of dump valves, and these are opened by an automatic control system either to maintain a set steam pressure or a fixed value of Tcold, depending on how you are using the system.

The advantage of the turbine bypass system is that the water is not being wasted; it can be pumped out of the condenser and then be re-used as feedwater. Turbine bypass will require your condenser to be under vacuum, and several turbine systems have to be up and running before it will work, so at your PWR this isn't considered to be a system with a true nuclear safety role, such as in responding to a fault. You might have to use the PORVs for that.

As you take your reactor critical and increase the heat input to the primary circuit, you'd expect to see the rate at which you are dumping steam increase to control Tcold: you have more heat to get rid of. To increase steam dumping requires either the PORVs or the turbine bypass valves to progressively open.

You'll see later how the transition is made from steam dumping to running a turbine, but you should already be able to appreciate that the reactor is hardly going to notice—it won't know where the steam is going!

12.6   And Finally… Boron

I've mentioned boron quite a few times in this book. Boron is a light element, with just 5 protons. A fifth of naturally occurring boron exists as boron-10 (with 5 neutrons), and the rest is boron-11 (with 6 neutrons). We won't worry too much about boron-11 here, but boron-10 is interesting because it fiercely captures thermal neutrons.

You can use boron to 'balance' the reactivity of the core as conditions within it change. Most importantly, by adding lots of boron to the primary circuit at the start of an operating 'Cycle', you can compensate for the very high reactivity of all of that new fuel. As the fuel is burnt up and its reactivity falls, you can slowly reduce (dilute) the dissolved boron so keeping reactivity balanced,

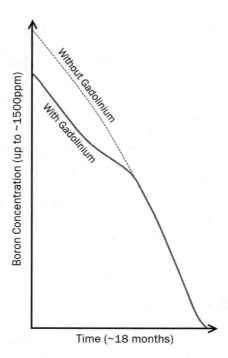

Fig. 12.5 Boron Letdown Curve

and the reactor critical at the programmed temperature. Figure 12.5 shows how the concentration of boron dissolved in the primary circuit is reduced over time (here in an 18-month Cycle). It's called the 'Boron Letdown Curve' because the dilution is performed gradually using the letdown flow of the CVCS.

Figure 12.5 is more of a curve than a straight line. This curve emerges for several reasons. Firstly the negative reactivity worth of each bit of boron is not the same—the more boron that is present in the primary circuit, the less affect a small change will have. Each part per million (ppm) of your dissolved boron might be worth around minus 6 milliNiles at the start of a cycle, rising to minus 8 milliNiles per ppm at the end of a cycle. Secondly, the change in core reactivity with burn-up is complicated by changes in power shape within the core.

The most dramatic deviation from a straight line is at the start of the cycle. It's caused by a deliberate poisoning of new fuel pellets with a material that captures neutrons: Gadolinium. In effect, your core designers are nobbling the fuel early in an operating cycle to reduce its starting reactivity. This might seem an odd thing to do, but it avoids very high boron concentrations, keeping the moderator temperature coefficient (MTC) from going positive.

Reducing the boron at the start of a cycle also improves the chemistry in the primary circuit. Boron in the primary circuit will be in the form of boric acid (BH_3O_3). This is what the chemists call a 'weak' acid, but even so, they'll need to add a small amount of alkaline, typically lithium hydroxide (LiOH), to balance the pH. If you had no gadolinium and even more boron, you'd need more LiOH added to the primary circuit. But, a very high level of dissolved lithium has been linked to corrosion processes in stainless steel pipework, so is best avoided.

Gadolinium is a 'burnable poison'. In other words it's used-up as it captures neutrons. This means that it only has a significant effect on the reactor in the first part of the cycle—it's doesn't cause a long-term reduction in your core's reactivity.

12.7   Routine Dilutions

Typically, you're going to be diluting the boron concentration by 2–3 ppm (parts per million) every day over the length of the cycle. How do you know when to do it? Well, think about what's happening in the reactor. As you burn-up the core from minute to minute and hour to hour, its reactivity will naturally fall giving a downward pressure on reactor power. But the reactor is following the steam demand. So it has to supply enough steam to match whatever the turbines are demanding and to do this it has to stay critical.

In practice, reactor temperature will fall slightly if more power is taken by the turbine than the reactor is providing. This temperature reduction will increase reactivity (through the MTC and FTC) and bring reactor power back-up to match the turbine but at a lower temperature. Over a few hours the reactor temperature, measured at, say, Tcold, will slowly drift downwards, perhaps by a few tenths of a degree C. If you are watching Tcold, you'll be able to see this, and you'll know it's time to make a small reduction in boron.

If you did nothing, the control rods will eventually (automatically) move out a step or two to restore Tcold to its programmed value; that's another way of reminding you that a small dilution is due. What do I mean by small? Perhaps a few tens of litres of demineralised water added to the primary circuit, two or three times per day near the beginning of the cycle, rising to a few hundred litres each time towards the end of the cycle. As the boron concentration drops, you will have to add progressively more fresh water to get the same ppm change in boron concentration. If that's not clear, imagine a primary circuit at 1000 ppm: if you replace a tenth of its water with fresh water, boron would fall by 100 ppm. If you start from only 500 ppm, that same amount of fresh water only gives you a 50 ppm reduction in boron, or roughly half the increase in reactivity.

Changing the boron concentration is important in other ways when controlling your reactor. Boron can be used to shutdown your reactor, or to keep it shutdown as you cool down the plant. If you add lots of boron, then no amount of cooldown will bring the reactor back to criticality, despite the negative temperature coefficients. Another way of saying this is to say that you can use boron to give you an adequate 'Shutdown Margin'. Boron can also be used to compensate for large, slow changes in reactivity that follow power changes, especially those caused by Xenon-135 which we'll meet in a later chapter.

You might ask what you're going to do when the boron has all been diluted out of the primary circuit? That's when you need to shut down the reactor: the cue for a 'Refuelling Outage'. We'll take your reactor through one of those in Chap. 21.

13

Putting a Spin on It

Back in Chap. 7, I guided you through taking your PWR critical and raising power up to a few percent of full power. You'll understand that the steam that your reactor is now producing is being 'dumped'; you're not doing anything useful with it… yet. In this chapter, you're going to drive your reactor a bit harder and start generating electricity.

13.1 Stable at Low Power

Remember that a few things happened to your plant as you raised it to low power. Let's look at them again:

- Thot has risen above Tcold. This shows you that you're adding enough heat to the water flowing through the core to see a temperature increase from bottom to top.
- The rate at which you're dumping steam has increased, and you've had to increase feedwater flow to maintain normal steam generator water levels. The more heat you put into your primary circuit, the more steam you're going to produce from your steam generators, and the more water you'll need to add to the SGs to replace the steam.
- The level of water in the pressuriser will rise. This is a bit more subtle. Before you started raising power and temperature, the average temperature in the primary circuit was pretty much the same as Tcold. As you increase power, Thot rises, so the average temperature in the primary circuit (roughly half the sum of Thot and Tcold), also rises. By the time you get to full

© Springer Nature Switzerland AG 2019
C. Tucker, *How to Drive a Nuclear Reactor*, Springer Praxis Books,
https://doi.org/10.1007/978-3-030-33876-3_13

power, the average temperature would have risen by around 16 °C. Even a small change in average temperature will have affected the density of the water in the primary circuit—it becomes less dense, so takes up more volume. The only place the less dense water can move into, to accommodate this increase in volume, is along the surge line and into the pressuriser. This means that the pressuriser water level will have risen as you've raised power. Your PWR has a pressuriser that's big enough to cope with water density changes all the way up to full power. That means that you don't have to add or remove water from the primary circuit, as you change power.

- The start-up rate (SUR) will have dropped back towards zero, without you moving the control rods. If you think back to the chapters describing the stability of the reactor, you'll see that both your fuel and moderator temperatures will have risen as you increased power. The negative reactivity effects from this temperature rise will reduce your (SUR) until it is back at zero. If you're not increasing power, temperatures and reactivity will balance out—they will find their own equilibrium.

13.2 Supporting Your Turbine

Up until now, the steam that your SGs have been producing has been 'dumped'. This could have been to the atmosphere (via the power operated relief valves, PORVs) but more likely you've been using the turbine bypass system to dump steam to the turbine condensers. This is much quieter… and it means that you can recover the water and re-use it as feedwater.

To dump steam to the turbine condensers, you've had to start-up a lot of the systems that would typically support the turbine's operation at power. As I've already mentioned, the condenser needs to be under vacuum to condense the steam at an efficient low temperature. Actually, it's the condensation of the steam that maintains that vacuum, but you still have to pump the air out to begin with, so 'Air Extraction Pumps' (vacuum pumps) will need to be used.

Your turbine will need lubrication, so expect to see some oil pumps and oil coolers in service. You'll also need a way of sealing the turbine shaft against the edges of its casing, so as not to lose steam or vacuum. Your turbine, like most others, does this by injecting a small amount of steam into special seals around the shaft—known to the engineers as 'Glands'. It's not dissimilar to the reactor coolant pump sealing water arrangement we saw on the primary circuit—some steam goes into the turbine through the gland, some comes out, but the effect is that you maintain the seal around the shaft.

Glands only work effectively if the turbine shaft is turning, so your turbine includes a small electric motor able to turn the shaft at low speed; known as 'Barring Gear'. Unfortunately, if you simply try to start your turbine spinning using the barring gear, nothing will happen. The turbine shaft with all of its blading weighs a few hundred tonnes, and even with the lubricating oil system in service, it's not going to move very easily. The designers of large turbines have an answer to this: 'Jacking Oil'. The jacking oil system pumps oil (the same oil as the lubricating oil) at high pressure, under the turbine shaft. This lifts the shaft off its bearings, and the barring gear can then start it turning at a few tens of revolutions per minute. Jacking oil is amazingly effective. If the covers are off of the turbine during maintenance, and you turn-on the jacking oil (with the bearings still intact), you can spin the whole turbine by hand!

Circulating the steam and feedwater through the SGs and turbine condensers allows you to control its chemistry. For example, as the steam passes into the condenser, dissolved gases will be released and then will be swept away by the air extraction pumps. Similarly, you can add chemicals such as ammonia and hydrazine to the feedwater; keeping it in a condition where pipework and SG corrosion is minimised.

At the end of the shaft, you have a generator. You can have a closer look at this later, but for now, I'll just mention that large generators are typically cooled with hydrogen gas and that your generator has a 'Seal Oil' system that needs to be in service to keep the hydrogen inside the generator.

Finally, there's a separate turbine oil system used to open the hydraulic stop and governor valves. As these valves run very hot, they have to use special oil, commonly known as 'Fire Resistant Fluid (FRF)'. Figure 13.1 shows all of the

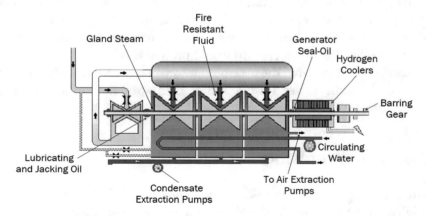

Fig. 13.1 Turbine support systems

turbine support systems that will need to have been put in service before you're ready to start spinning the turbine using steam. With any luck, this was being done while you were driving your reactor up to low power, so it'll be waiting for you…

13.3   Spinning-Up

With your turbine shaft spinning slowly and its supporting systems now in service, all you need to do is to fully open the stop valves on the turbine and then just open the governor valves by a tiny amount. Admitting less than half a percent of full power steam to the turbine is all that's needed to see it accelerating away. You'll need to take it from barring speed (a few tens of rpm) up all the way up to its normal speed of 1500 rpm. As you do this, you might see the steam dump valves close in a little, as you'll be sending some of your steam through the turbine blades, but there will be almost no effect on the reactor.

You can raise turbine speed by a few hundred rpm every minute, but as the steam passes through the blades the turbine will heat-up, and this can cause problems with vibration and with reduced clearances inside the turbine. These are huge machines, and typically there will be speed ranges that you'll need to move through slowly and others where you'll move more quickly—follow your procedures!

Once you get the turbine up to full speed, and if you haven't run the turbine for a while, you may be asked to perform an 'Overspeed Test'. If you raise the turbine speed a bit further (say, by another 10% or so), the turbine should 'trip' with the stop and governor valves quickly closing to avoid damaging the turbine from it spinning too fast. Overspeed tripping of your turbine is really important: if it was ever suddenly disconnected from the grid but with stop and governor valves still open, it would accelerate until it failed—probably catastrophically. The overspeed trip would prevent this.

When you've completed the overspeed test, you can reset the trip, open the stop and governor valves (just a little), then spin the turbine back up to 1500 rpm.

13.4   Synchronising

You're still not generating any electricity, so now's the time to look a bit more carefully at the generator. As I've already mentioned, your generator is in two main parts: the 'Stator' (because it doesn't spin, i.e. stationary) and the 'Rotor' (the bit that spins). Both of these are wrapped with copper bars in a complicated arrangement. The stator will have water flowing in between some of the copper bars to provide cooling. This isn't possible on the rotor, so hydrogen gas is blown over the rotor then through coolers, to take away any heat. Why hydrogen? Because it's physical properties are great for removing heat, and it's cheap. Helium would be good for this too, especially as it's non-flammable, but it's far too expensive.

The rotor has an electrical current running through it that produces a magnetic field. This current is produced in a smaller generator—known as an 'Exciter'—attached to the same turbine shaft. Typically on larger turbines like yours, two stages of exciter are needed: a small permanent-magnet generator (the 'Pilot Exciter') generates a magnetic field for a more substantial 'Main Exciter', which in turn generates the current for the generator rotor.

All of this makes the generator rotor both complicated and large. In Fig. 13.2, you can see an example of a generator rotor (from a 600 MW turbine), being inspected before being lifted into the Generator. This one weighs 70 tonnes.

So, to recap: you now have turbine blades spinning at 1500 rpm. These are coupled to your generator rotor, so it is also spinning. When you turn-on your exciter, your rotor will see a large electrical current and from this will form a

Fig. 13.2 Generator Rotor

spinning magnetic field. However, your stator is not yet connected to the grid, so you can't be sending any power offsite.

There's a fast-acting switch ('Circuit Breaker') between your generator stator and the transformers leading to the grid, but if you simply close it, things are likely to go very badly. The chances are that your turbine speed and the grid frequency won't be quite the same. Even if they are, they may not be aligned ('in-phase'). Remember, your generator produces 3-phase AC electricity at 50 Hz, and at around 20,000 V. If you managed to close the generator circuit breaker out of phase or at a different frequency from the grid, the electrical forces acting on the shaft would be enormous—there's every chance that your turbine would be wrecked... along with the building it's sitting in.

To avoid this, you're going to use a synchroscope. It's a simple device: it measures the frequency and alignment (phase) of the current being produced in the generator rotor, together with the frequency and phase of the incoming grid supply, measured on the grid side of the circuit breaker. The synchroscope displays the difference between the two on a spinning dial (Fig. 13.3). It won't let you close the circuit breaker until this all matches, so you need to keep adjusting the turbine speed and phase (by changing how much steam passes through the blades) until you get a good fit—shown by the dial being stationary and (usually) pointing straight up. Once you're at the correct speed and in-phase with the grid, the circuit breaker switch can be closed to connect the generator stator to the Grid.

From now on your turbine will always stay electrically locked to the grid. If the grid speed (frequency) changes, even by a little, your turbine speed will do the same. The magnetic field produced by the rotor and the one in the stator push against one another and won't allow your turbine generator to move at a

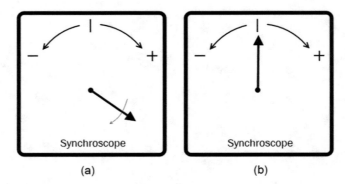

Fig. 13.3 Synchroscope. (a) Moving pointer. Indicates turbine is spinning at a different speed (frequency) to the grid. (b) Stationary pointer. Frequency is the same. Vertical position confirms that turbine is 'in-phase' with the grid

different frequency from the grid, nor to slip out of phase with it. This means that you can't simply speed up your turbine to generate more electricity. So how do you put more energy (electricity) into the grid?

You do it by allowing more steam to pass through the turbine blades *trying* to push the shaft and the rotor around faster. What you'll actually be doing is increasing the torque on the shaft leading to the generator rotor. The magnetic field produced by the rotor will then push against the magnetic field in the stator, and that will push against the grid, to try to make all of the other generators and equipment attached to the grid spin just a little bit faster. It's a big grid, so you'll not see any measurable change, but you will be putting energy in, just by trying.

Incidentally, this is also true of almost every other generator on the grid. All the other turbines, all over the country, will be spinning virtually in phase with yours and at the same speed (or a multiple of it, depending on how they are wired). Most designs of wind turbines spin synchronised to the grid; which is why when you look at a wind farm, all of the turbines are spinning at the same speed, with the blades in the same positions (any that aren't are probably not connected to the grid). Just like your turbine, the wind turbines are locked to the grid frequency and in-phase with it because of the magnetic fields in their generators.

If all of the generators on the grid are putting too much energy in, compared with the electricity demand, the grid frequency will rise. If they put too little in, the grid frequency will drop. It's the job of the grid controllers to balance the supply from the generators with the total demand on the grid, to keep the frequency close to 50 Hz at all times—it's not an easy job, as you saw how variable the electricity demand can be in a previous chapter.

13.5   Turbine Power Raising

When you first synchronised your turbine to the grid, the governor valves will have allowed just enough steam into the turbine to generate a few tens of MW of electricity. It's not a lot compared with your full power of 1200 MW, but it will ensure that your turbine is continuously pushing against the grid. If you didn't do this, the grid could drive the generator as if it was an electric motor—and this could lead to overheating.

Your steam dump valves will have closed a little to compensate for the steam that's now being used to drive the turbine, but you're probably still dumping steam. To close the dump valves, raise turbine power (by opening the governor valves a bit) until all of the steam being produced in your SGs is being used to drive the turbine. Your dump valves will have closed. Now your reactor and turbine are balanced at low power, and you're finally getting paid (a little) for electricity.

The next step is to raise both reactor power and turbine power up to your full power limits. This isn't quick. Just as when you were spinning your turbine up to 1500 rpm, your turbine could be damaged if you try to change its power output too quickly. Some turbines are better at faster power changes than others. Yours is typical, able to increase power at just a few MW per minute. This means it's going to take you somewhere between 5 and 10 h to reach full power, assuming that there aren't any limits on the reactor that would force you to move more slowly.

Modern turbine governor valves are controlled by computers. Once the turbine is synchronised to the grid, you only need to type in a target load (in MW) and a rate at which you want to get there, and the computer will do the rest, monitoring vibrations, expansions and any other concerns as it goes along.

You've already learnt that your PWR will follow steam demand, so it looks like all you have to do to raise the power on the reactor is to tell the computer operating the turbine to increase its output? Unfortunately, it's a bit more complicated than that, as you'll see in the next chapter.

14

Going Up!

14.1 Reactor Power Raising

Let's remind ourselves what happens when you open the governor valves on your turbine. You admit more steam into the turbine, so you're going to be pushing the blades, shaft and generator rotor a little bit harder. This will push against the grid frequency, so you'll be putting more energy (electricity) into the grid. All well and good, that's probably what you wanted. But what's happened to your reactor?

By taking more steam from your SGs, you will have caused a drop in steam pressure. This creates a drop in steam temperature (Tsteam) which in turn causes a reduction in the temperature of the water heading back to the reactor (Tcold). A decrease in Tcold reduces both fuel and moderator temperature, resulting in positive reactivity feedback. The reactor power will start to rise until it matches the steam demand at which point a new balance in power and temperature will have been achieved. As the reactor driver, you haven't needed to do anything.

But your reactor is now operating at the 'wrong' temperature. It's off-programme. In this scenario, it will have ended up colder than you want it. Your steam temperature and pressure won't be at their designed values, and that will prevent the turbine from working quite so efficiently. What you need to do is to raise the reactor temperature up a little, and you can do that by withdrawing your control rods a few steps. This increases reactivity, allowing temperatures to rise (negative reactivity) to achieve a new balance—this time at the right temperature. You can let the control systems do this automatically, or step in and do it manually, using the control rod controls in front of you.

© Springer Nature Switzerland AG 2019
C. Tucker, *How to Drive a Nuclear Reactor*, Springer Praxis Books,
https://doi.org/10.1007/978-3-030-33876-3_14

14.2   Power Defect

That's all very well for a small change in power (up or down) but what about more substantial changes, such as moving from very low power all the way up to full power? The answer is that control rods aren't the way to control such a large change in power.

On a PWR, you will always lose reactivity if you increase power. What I mean is this: if you increase power, both fuel and moderator temperatures will rise. Each of these will result in negative reactivity, as you've seen in earlier chapters that described the Moderator Temperature Coefficient (MTC) and Fuel Temperature Coefficient (FTC). You can plot a graph of these two effects added together. It's called a 'Power Defect' graph (Fig. 14.1).

Figure 14.1 shows how much reactivity you will lose as you increase power from zero power up to full power, assuming that you keep the reactor on the correct temperature programme. As you've already seen, the MTC depends on the boron concentration in the primary circuit, which depends on how far your PWR is through its operating cycle. So in reality, there's a different power defect graph, for each point in the cycle. Figure 14.1 is a graph for about mid-way through. You can see that it's made-up of contributions from both the MTC and the FTC, with the FTC contributing a little more at this time. By the end of the cycle, the MTC's contribution will be the larger of the two.

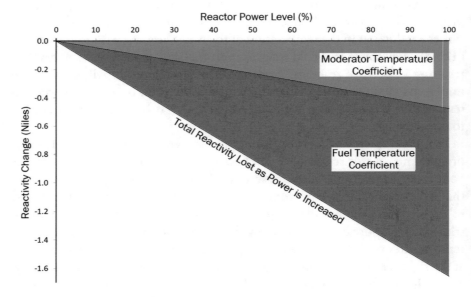

Fig. 14.1 Mid-cycle power defect

Look at the scale of Fig. 14.1. It's in Niles (see Chap. 3), so the power defect has a significant effect on reactivity. All of the control rods added together are stronger than this (around 8 Niles), but this wouldn't be true if you just considered the control banks. Remember that the shutdown banks are only used in shutting down the reactor; they will be fully withdrawn with your reactor critical, so you can't vary their position. Withdrawing the control banks on their own couldn't add enough reactivity to overcome the power defect unless they were very deeply inserted at low power. But there's problem with that idea: control rods distort the shape of the power in the reactor.

14.3   Power Shape

Figure 14.2a shows a full-power shape for the power in your PWR (top to bottom). This power shape is for part-way through an operating cycle, with your control rods nearly all out of the reactor. The dips in the trace show you where the gridstraps are on your fuel assemblies (Chap. 3), as gridstraps capture neutrons. There's a slight bias in the power shape towards the bottom of the core. This is because the lower half of the reactor is colder—and therefore more reactive—than the top. The reactor physicists use two terms to describe this, the 'Axial Offset', or the 'Axial Flux Difference (AFD)', but both of them are simply measures of the percentage difference in power between the top and bottom halves of the core.

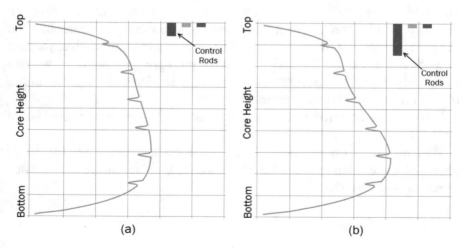

Fig. 14.2 Axial power shapes with control rods out, and with control rods inserted

Compare Fig. 14.2a with Fig. 14.2b. In Fig. 14.2b, the control banks are more deeply inserted into the reactor (from the top). Reactivity has been balanced by a change in boron concentration, so your reactor is running at the same power level, but look how much more the power shape is biased towards the bottom of the core. Pushing the rods in distorts the power shape, even though the overall power level is unchanged. This means that some of your fuel pellets (in the lower part of the core) are now producing much more power than they were with an undistorted power shape, so will be burnt-up faster. More significantly, it also means that the decay heat will be biased towards those same fuel pellets.

Concentrating the power in one part of the core makes fuel failure more likely if a reactor fault were then to occur. This is why you try to run your PWR 'Rods Out', with an AFD near zero, reflecting an even spread of power between the top and bottom of the core.

So, practically, how do you raise power if you can't use control rods to compensate for the power defect? You dilute, reducing the boron concentration in the primary circuit. As soon as you start raising the power on the turbine (or even a little before), you start adding fresh (unborated) water to the primary circuit. This compensates for the reactivity you're losing through the power defect as reactor power rises. You continue to dilute throughout the power raise, either by monitoring Tcold and AFD or by following a dilution plan provided by a computer model of the core.

As you add fresh water through the Chemical and Volume Control System (CVCS), excess primary circuit water will be diverted to your 'Radioactive Waste' plant for processing. You could work out how much of a dilution you need using tables and graphs if you know enough about your core, but you can probably see that the computer will be quicker and more reliable. You aim to reach full power, on the designed temperature programme, with the control rods nearly fully withdrawn—and that may require an overall dilution of tens of tonnes of unborated water.

Once you've achieved full reactor and turbine power, everything should be stable?

Well no, unfortunately not. There's one more significant bit of reactor physics that you need to understand to successfully drive your reactor. It's due to xenon, and it's a doozy…

14.4   Iodine and Xenon

Around 6% of fissions of uranium-235 result in the fission product iodine-135 (I-135). Iodine-135 decays (by beta decay) to xenon-135 (Xe-135). Normally, Xe-135 would decay to caesium-135, which is nearly stable. Aside from the radiation released, this is pretty dull, and none of this is going to have a direct effect on your reactor. However, Xe-135 is also voracious as far as neutron capture is concerned. It captures lots of them! So sometimes a Xe-135 atom will capture a neutron (forming stable Xenon-136), rather than decaying. This is shown pictorially in Fig. 14.3:

Any xenon-135 you have in your reactor will capture some of your neutrons so will clearly give you a negative reactivity contribution. Xenon-135 is a reactor 'poison'. But it's xenon's behaviour during power changes that makes it important and interesting to you, as you drive your PWR. I-135 has a half-life of six and a half hours. For Xe-135 it's just over 9 h. These timescales are short enough to have a visible effect on how you drive your reactor, without being so short that they'd only be of interest to physicists.

14.5   Xenon Build-Up

Let's start with a reactor that's been shut down for a few days. During that time, all of the I-135 and Xe-135 that might have been produced in previous operation will have decayed away. Your reactor is xenon-free. Now you start-up

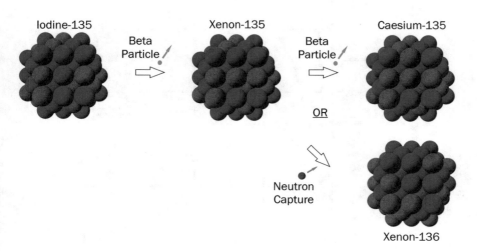

Fig. 14.3   Xenon-135 production and removal

your reactor… you'll begin producing I-135 as a fission product. Over the next few hours some of the I-135 will decay to Xe-135. As you raise power the rate at which you produce I-135 will increase, followed by an increase in the rate of Xe-135 production. But remember, both I-135 and Xe-135 will decay away with time, so you're going to reach an equilibrium level for each, where the production matches the decay. In fact, things are a little more complicated for Xe-135, as its removal is made up partly from decay and partly from neutron capture (to Xe-136). We'll come back to that later.

Figure 14.4 shows you a graph of how iodine-135 and xenon-135 concentrations in your reactor change with time following an unrealistically fast (!) increase in power from zero to full power. You'll see that both I-135 and Xe-135 reach stable levels with the Xe-135 lagging behind the I-135. Remember: it's only the Xe-135 that has an effect on reactivity so what Fig. 14.4 is telling you is that even after you've reached a stable power level in your reactor, the reactivity will still be changing. As xenon builds-in over the next 2–3 days, you are going to have to continue to dilute (albeit at a lesser rate than when you were power raising) to offset the negative reactivity appearing from the xenon.

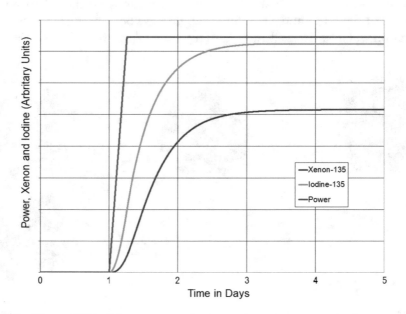

Fig. 14.4 Xenon-135 build-up

14.6   Xenon After a Trip

Xenon's behaviour after a trip is even more of a problem. You might imagine that when you trip, xenon will simply decay away but that's not what happens. It actually increases! At the moment you trip your reactor, your equilibrium level of Iodine-135 corresponds to the reactor power you had before you tripped. I-135 will decay away from this level with a simple 6.5-h half-life. However, this decay is the *production* route for xenon-135. So, to begin with, you'll be producing Xe-135 at the same rate as you were before the trip.

Now think about the processes that remove Xe-135. The natural decay to caesium-135 will continue as before, but this was only half the picture. Your equilibrium Xe-135 level was also due to the removal of Xe-135 by neutron capture. But there aren't any (or hardly any) neutrons any more as you've just tripped your reactor, so this removal mechanism has stopped. Your reactor is now producing more Xe-135 than it is losing, so the level of Xe-135 and its negative effect on reactivity goes up.

This is shown in Fig. 14.5. You can see that the result is dramatic, significantly increasing the negative reactivity effect from the Xe-135. It also lasts a while, your reactor won't be back to where it started (in terms of Xe-135 negative reactivity) for around 20 h, after which you'll gradually see an improvement as the Xe-135 decays away.

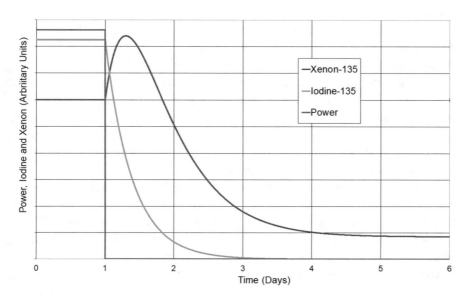

Fig. 14.5   Xenon-135 after a reactor trip

On early reactors such as the MAGNOX plants (Chap. 22), the build-up of xenon after a trip meant that you couldn't achieve criticality for around 24 h post-trip, even with all of the control rods fully withdrawn. On a PWR, for most of an operating cycle, you can recover reactivity by reducing the boron concentration in the primary circuit. Even so, if you're trying to start-up your reactor within a few days of a trip, predicting the point of criticality has to take the xenon-135 transient into account. It's going to be a moving target.

14.7   January Sales

Some people struggle with the behaviour of xenon when they start working with reactors—I know I did! I've seen several attempts at explaining it, for example, with diagrams of interconnected bathtubs—none of which really worked for me. So, let me try something completely different…

Imagine a West End store during January Sales. There are hundreds of people walking around the store, and more queued up outside. Some of the people inside get fed-up and walk out without buying anything. Others find a bargain, buy it, and then leave. This can all balance out quite nicely, with the people coming in matching those leaving. In this analogy, the people in the store are your Xenon-135. The people outside in the queue are the iodine-135, and the bargains represent the neutron flux (power) of your reactor.

So what happens if you trip? In our analogy, it's as if all of the bargains suddenly disappear from the shelves, so no-one will be able to buy anything. People will still leave the store, but some of them will still be wandering around looking for those elusive bargains, so their leaving won't speed-up appreciably. It also takes time for the message about the missing bargains to reach the queue outside. So to begin with, people will still be flooding into the store just as if it were fully stocked. Just like in your reactor, even though the bargains have disappeared (the flux has gone to zero), the number of people in the store (the xenon-135) will actually go up.

Eventually, the story about the missing bargains will spread, and people will stop joining the queue outside (no more iodine produced). As the pressure in the queue falls, the number of people entering the store will drop below the number leaving, and the total number of people (xenon-135) will fall, reaching zero when everyone's finally gone home.

You can use the same analogy for smaller decreases or increases in power (changes in the supply of bargains). For example, if the number of bargains on the shelves suddenly increases (power increase) the number of people buying and then leaving will also increase. However, the people in the queue take time to realise the change, so won't initially be trying to get into the store any quicker. This will mean more people leaving than entering and an overall fall in the total people in the store (xenon-135). And so on…

It's not physics, but it might help?

15

Power, and How to Change It

15.1 The Toolkit

You've now been introduced to all of the tools you need to drive your reactor.

To begin with, you needed to understand the concept of reactivity ("how friendly your reactor is to neutrons").

This enabled you to see how changes in fuel and moderator temperature in a Pressurised Water Reactor (PWR) feed-back as changes in reactivity and power. As this is negative feedback, you've seen how this keeps a PWR reactor stable with little or no intervention from you, the operator.

Along the way, you've seen how reactor power is measured (or calculated), and how quickly your reactor can be shut down; although you can't just turn off the 'decay heat'.

As part of the start-up procedure, you've seen what happens when you connect your PWR to its turbine. Again, negative feedback effects—in this case, due to changes in steam generator pressure and temperature—have a strong stabilising effect on the plant and allow the reactor to follow the steam-demand.

Finally, you've seen the effects on reactivity from dissolved boron, control rods, power defect and xenon-135. The first two of these are under your direct control; the other two are a consequence of what you do with reactor power.

© Springer Nature Switzerland AG 2019
C. Tucker, *How to Drive a Nuclear Reactor,* Springer Praxis Books,
https://doi.org/10.1007/978-3-030-33876-3_15

15.2 Practical Example: A Significant Power Reduction

Imagine you've been running your reactor at full power for some weeks. You're in the middle of an operating cycle. Everything is stable, there is little or no control rod movement and you just need to do the occasional dilution of the primary circuit to keep Tcold on its programme.

Then you get a phone call: "There's a problem with the turbine; you need to drop to 75% power?"

What do you do?

Well clearly the big red button isn't going to help you—that would shut down the plant completely.

You might be tempted to rush to the reactor control panels and start driving in the control rods. But by now you'll probably appreciate that driving the rods into the core will only cause a temporary drop in power, as you'll still be taking full steam for the turbine. The plant would cool-down, and power would simply return to 100%—but well off its intended temperature program.

So let's be sensible. If you need to reduce plant power, you need to start at the turbine. Use the computer that controls the turbine governor valves to slowly close them in. As you do this, less steam is admitted to the turbine, and the electrical output will fall. Ideally, you'd do this at just a few MW per minute, but if you're in a hurry, you can reduce output by tens of MW per minute if you're not making too big a change in power overall.

As the governor valves start to close in, the steam flow path becomes more restricted so the upstream steam pressure will rise. This means that the steam generator steam pressure will increase, taking the secondary side temperature and then Tcold along with it. Now reactor power will start to fall as the fuel and moderator temperature coefficients (or the power defect, if you prefer) reduce reactivity. Your reactor will follow the turbine as the turbine power is reduced. As Tcold moves away from its programme, your control rods will step in, providing additional negative reactivity and ensuring that Tcold doesn't rise too far.

You're now driving your reactor by moving your turbine power. You've already seen this in Chap. 12, but there's a problem. While these feedback mechanisms work fine for a small change in power, you've been asked for a whole 25% reduction. If you drove your plant that far just using the turbine, you'd end up with the rods very deeply inserted in the core. This distorts the power shape so probably isn't allowed by your procedures. On some plants, there are even limits built into the automatic rod control system that would prevent the rods from driving in that far, in which case you'd reach a very high Tcold as power was reduced.

I expect you've already guessed that you can overcome these problems by adjusting the amount of boron in the primary circuit. So let's go back to that phone call…

15.3   What You Really Do

Instead of reaching for the turbine controls straightaway, have a look at the example power defect graph in the last chapter. If you're dropping power from 100% to 75%, the Power defect will change by about 0.4 Niles (i.e. from −1.6 Niles to −1.2 Niles), or plus 400 milliNiles. Remember that Power Defect is always a negative contribution to reactivity, so if it is reducing in size, that means that overall core reactivity is increasing. If you want to offset this positive change in reactivity by changing the level of dissolved boron, you're going to need to increase the boron by around 60 ppm (roughly minus 400 milliNiles divided by minus 7 milliNiles per ppm—the reactivity worth of each ppm of boron).

So here's what you really do:

- Go to the control panel for the Chemical and Volume Control System (CVCS) and dial in a 'boration' (adding boron) equivalent to a boron increase of 60 ppm; your procedure will tell you how to calculate this for your plant at the current burn-up. Now, set-up the CVCS controls so that this boration happens gradually over the same timescale in which you want your power reduction.
- Set-up the power-reduction on the turbine governor controls but don't start the valves moving just yet. The pipework from the CVCS to the primary circuit is quite long, so wait a few minutes until you see Tcold just start to fall—telling you that the primary circuit boron has started to increase—then begin the turbine power reduction.

What you will now see is that the reactor and turbine power will reduce together, with Tcold staying on programme and rods hardly moving at all.

15.4   Controlling Axial Power Shape

You can't leave things entirely to the automatic systems though. As you reduce the power the top of the core cools down a bit more than the bottom of the core, increasing the reactivity in the top relative to that in the bottom. This causes the power shape to naturally shift upwards, giving you a more positive Axial Flux Difference (AFD). There are tight procedural limits on AFD at most stations, so this positive movement in AFD can't be left unchecked.

As the power reduction proceeds, keep an eye on AFD. If it's moving too far positive, pause the boration for a few minutes. With less boron being added to the primary circuit, temperatures will rise as turbine power falls. As Tcold rises, your control rods will automatically drive inwards to hold Tcold on target. As the rods drive in, the axial power shape (and hence AFD) is pushed downwards—more negative. If you're lucky, you'll have a computer program telling you how much of a pause in boration you need in order to keep AFD within limits. If not, you'll just have to carefully monitor the plant. Lots of practice on the simulator will help—and don't worry, your training department *will* have simulators!

15.5   And Xenon

Once you've achieved 75% power on both your reactor and turbine, the job's all done? No, unfortunately not. The change in reactor power will kick-off a transient change in your xenon-135 level. If you drop reactor power, it's a bit like a mini-trip, as you can see from Fig. 15.1.

Just like after a trip, if you reduce reactor power, your equilibrium level of iodine-135 corresponds to the reactor power you had just before the reduction. I-135 will decay away from this level towards a new (lower) equilibrium level, but its 6.5-h half-life means that this isn't an instant change. Once again, the decay of I-135 is the *production* route for xenon-135. So, to begin with, you'll be producing Xe-135 at the same rate as you were before the reduction.

The natural decay of Xe-135 will continue as before, but there are now fewer (only 75%) of the neutrons flying around compared with being at 100% power. This means that the removal of Xe-135 by neutron capture will reduce. So just like after a trip, your reactor will be producing more Xe-135 than it is losing, and the level of Xe-135 and its negative effect on reactivity initially goes up.

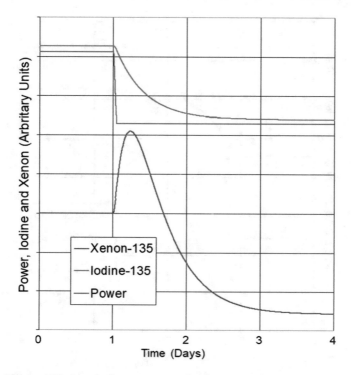

Fig. 15.1 Xenon-135 change for a power reduction

Unlike a reactor trip, this isn't the end of the story. The Xe-135 will fall and settle out at a newer lower level. How long this takes, and how far Xe-135 will have initially risen, will depend on how fast and how substantial the power reduction was. As the reactor operator, you're going to have to ride-out this 'Xenon Transient' over the 10's of hours following the power reduction. In practice, that means that as soon as you've finished borating for the power reduction, you'll have to start diluting for the rising xenon. A few hours later, and you'll be borating again until everything levels out.

And you'll have to go through the whole thing again, in reverse, when it comes to going back to 100% power, this time with an initial fall in xenon (production being less than removal as power rises), followed by a longer-term rise (Fig. 15.2).

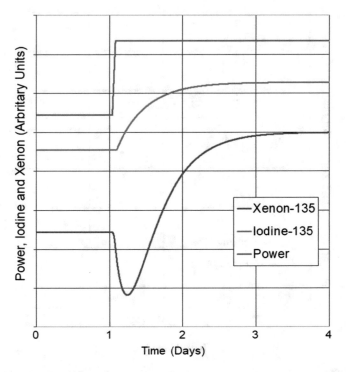

Fig. 15.2 Xenon-135 change for a power increase

15.6   Flexible Operation

Your PWR is easiest to drive if you run at steady full power. Most PWRs in the world operate in this way (as 'baseload' generators) because they are embedded in electricity grids where they contribute only a small part of the supply. This allows other types of power station that can more easily change their output—such as coal, hydroelectric and combined cycle gas turbines—to be used to match the supply and demand on the grid, from minute to minute and hour to hour.

In the UK, with only the occasional exception, the nuclear stations have always operated as baseload generators. This comfortable position (for the nuclear reactor operators) is threatened by recent changes in generators attached to the UK grid. Many of the large coal-fired generators have shut down permanently, and the remainder have limited lives. Gas has become more expensive, and there are now significantly more renewable, but less controllable, generating plants on the system (i.e. wind and solar). This could mean that nuclear plants will be forced—either physically or commercially—to operate more flexibly, rather than as baseload generators.

Luckily, there's a strong precedent for this: France. France has more than 50 PWRs, generating nearly 80% of its electricity. While they can export lots of electricity to their neighbours (including the UK), they can't keep all of their stations running at full power all of the time. Some of them have to operate flexibly.

At the simplest level, this means that some of the PWRs in France will be asked by their grid control to reduce power, perhaps by 50%, for a few hours at a time each day (or more probably overnight). The manoeuvre they will follow to achieve this would be the same one you've just been through in the text above. More drastically, perhaps at weekends when electricity demand is at its lowest, some of the French plants may be asked to shut down entirely for a couple of days, then restart. None of this is difficult on a PWR, especially if a few hours' notice is given. However, it will increase radioactive waste arisings (due to the extra borations and dilutions), and there will be more wear-and-tear on the plant as it undergoes temperature and pressure changes.

PWR power changes tend to be more difficult at the very start or end of an operating cycle. At the beginning of a cycle, the fuel will not have settled down, so it will be more prone to failure if power is changed rapidly. At the end of a cycle, relatively large volumes of water are needed for dilutions (the same change in ppm of boron takes more water when boron is starting from a lower level). However, if your fleet of PWRs is large enough (as in France), the operators can choose which of the plants to operate flexibly and which to use as baseload generators, thereby avoiding these problems.

Some of the French PWRs have a design feature that facilitates flexible operation: 'Grey Rods'. I've described how the control rods (RCCAs) in your PWR are made from a mixture of silver, indium and cadmium (Ag-In-Cd). This, together with their size and geometry, gives them the appearance of a 'black absorber' as far as neutrons are concerned. In other words, any neutron that enters into one of these control rods will be captured (the analogy being a 'black' surface, capturing, or absorbing light).

On some of the French PWRs, a proportion of the control rods have the Ag-In-Cd in some of their rodlets replaced with stainless steel. As far as neutrons are concerned, these rodlets are only 'grey', allowing some neutrons to escape. Why would you design control rods like this? Because they can be inserted deeply into the core quickly and easily, reducing reactivity to an extent, but without significantly distorting the power shape. In other words, grey rods can be used in power changes instead of—or to reduce—dilutions and borations.

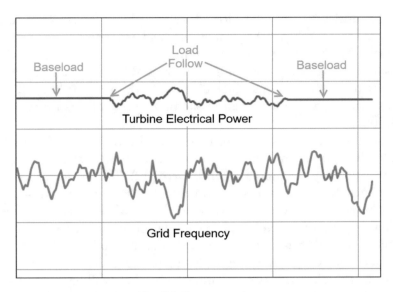

Fig. 15.3 Turbine baseload and load follow operation

15.7   Load Following

In the UK, Sizewell B has no grey rods, so it can only change power levels using its black rods and boron changes as described above. However, limited flexible operation of Sizewell B is designed-in through 'Load-Follow' control in the turbine governors. If requested, these will respond to small changes in grid frequency, with reactor power following turbine power over a (relatively) low range. As frequency rises, the governor valves close-in a little, as it falls, they open-up. Turbine power is then changing in a way that tends to flatten out the grid frequency on a minute-to-minute timescale (Fig. 15.3). The limited range means that there is no significant wear-and-tear on the plant.

The more modern PWRs being constructed in the UK—such as those at Hinkley Point 'C'—will typically include grey rods to more easily offer flexible generation as the mix of generators on the grid continues to change.

15.8   Taking the Long View

Operating flexibly, changing power, or shutting down and starting up the plant more frequently will all make your job as a reactor operator that much busier. The need to do any of these things will come from outside; from changes in the mixture of generators on modern-day electricity grids, or from commercial pressures. A trainee reactor operator 20 years ago probably didn't expect to be doing very much of this. The lesson here is that if you're planning to run your PWR for 60–80 years, you can't expect everything to remain the same, or be used in the same way, as the day you designed it.

Let me give you an example: At the MAGNOX stations (Chap. 22) it was common practice to shut down the reactors, from nearly full-power, by pressing the reactor trip buttons. At each station these were probably used once or twice per year (I was once allowed to trip a reactor from 900 MW, in just this way). But not long ago, I met one of the designers of the MAGNOX plants who asked me whether the 'Emergency Shut Down Buttons' had ever been used…? He was a bit surprised when I explained.

16

Steady Power with Nothing to Do?

16.1 The 'Q' Word?

So, your reactor is running at full power. Temperatures are on target, and xenon has reached equilibrium. Your power station is producing more than 1200 MW of electricity; that's about 3% of UK average demand.

Just how *quiet* is it in the control room?

Let's take a moment to think about how much automation is likely to be involved in driving your reactor.

Firstly, as you've already seen, your PWR is inherently stable due to temperature feedback effects. This leads to the 'reactor following the steam demand'. So, if the steam taken by your turbine is constant, then that will hold your reactor power at a constant value; I'll come back to this, as it's not as simple as it sounds, but your first guess might be that you're not going to need much automation at all.

16.2 Burning-Up

Your reactor is changing minute-to-minute, hour-to-hour as you're burning-up your fuel. Uranium-235 will be being lost from the reactor as it fissions. Some plutonium-239 will be being created, but some of that will fission too, so the number of fissionable nuclei will be falling. More significantly, your fuel pellets will be accumulating fission products, some of which (such as xenon-135) capture neutrons. When you consider all of these effects it should be clear that the reactivity of the fuel will be falling with time, but the reactor

© Springer Nature Switzerland AG 2019
C. Tucker, *How to Drive a Nuclear Reactor*, Springer Praxis Books,
https://doi.org/10.1007/978-3-030-33876-3_16

is still critical—it's running at full power to match the steam demand—so what else is changing to keep reactivity from going negative overall? The answer is temperature. If you do nothing, the reactor temperature (Tcold and Thot) will slowly fall. This slow temperature reduction gives you back some reactivity—just enough to keep the plant running—and I do mean slow; over a day, the drop in temperature might be less than one degree.

Your reactor and the rest of the power station are designed to operate at a specific set of temperatures and pressures for maximum efficiency. Your PWR has an automatic system to correct this fall in temperature; it'll move the control rods. If the control banks are withdrawn just a step or two, reactivity will be added to the core and temperatures will return to their programmed values. For you, the operator, the control rod movement—probably indicated by an alarm or some other audible or visual clue—will be the indication that reactivity has shifted.

The automatic movement of the control rods—the system that you could call a 'Reactor Temperature Control System (RTCS)'—is the first of the automated control systems that make your life as an operator that little bit easier (Fig. 16.1). But remember, the RTCS will operate during transients and faults as well as at steady power. For some faults this reduces their severity, in others, the RTCS can make things worse!

Unfortunately, a movement of rods out of the core will also result in a shift of power towards the top of the core. Axial Flux Difference (AFD) will become more positive and might eventually move outside of limits. There's also the problem that you're probably starting with the control rods pretty much out of the core anyway, so there's not much scope for further withdrawal. You need another way of gaining some positive reactivity, and of course you have it in the ability to reduce the amount of boron dissolved in the primary circuit.

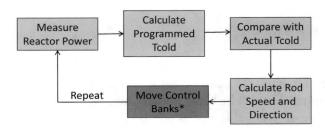

* NB: Control Rods move OUT if Tcold is below programmed value, and IN if Tcold is above.

Fig. 16.1   Reactor temperature control system (RTCS)

This is done through the chemical and volume control system (CVCS) as you've seen in earlier chapters.

The boron 'letdown curve' in Chap. 12 showed you how boron is reduced through an operating cycle to compensate for the falling reactivity of the fuel. An average reduction in boron of 2–3 ppm per day is typical, with the operator initiating dilutions a few times per day. This increases towards the end of the cycle when progressively more fresh water needs to be added to reduce an already low boron concentration.

Typically, CVCS systems are only semi-automated at PWRs. If they were fully automated and then malfunctioned, there'd be a significant risk of them causing a large, unwanted reactivity change and a plant trip. Instead, to dilute out some of the boron in the primary circuit, you'll need to set-up a desired volume (or 'batch') of fresh water for the CVCS to inject, then set it running, while monitoring that pumps and valves all work as intended. As the water is pumped into the primary circuit you'll see temperatures recover, and you'll be able to check that the dilution stops when it's supposed to.

16.3   Primary Circuit

There are two other primary circuit parameters that you'll be responsible for as an operator: pressure and water level.

You'll remember that in the pressuriser there's a steam bubble, with a level of water below that. So what determines the height of this water level? Many people guess (wrongly) that it's affected by the primary circuit pressure. As it happens, liquid water is pretty close to being incompressible, even at these temperatures and pressures, so small variations in primary circuit pressure won't show up as water-level changes. Instead, there are two other things that determine how the level behaves—primary circuit temperature and the CVCS flow balance.

The temperature of the water in the pressuriser itself is pretty constant at around 345 °C; the boiling temperature of water at 155 bar. But the temperature of the water in the rest of the Primary Circuit is linked very closely with reactor power. At zero power, with just decay heat in the core, the average water temperature in the primary circuit will be little more than 290 °C. On the other hand, at full reactor power with a Tcold of around 290 °C and a Thot of 325 °C, the average temperature will have risen to nearly 310 °C. While water is almost incompressible with pressure, it will still expand and contract with temperature changes. Where does it expand to as you raise temperature? It pushes up into the pressuriser (travelling along the surge line). So the pressuriser water level will naturally increase as you raise power and fall as you lower power (or trip). This is a significant effect, perhaps changing water level over half the height of your pressuriser, depending on its size.

I said that the second thing affecting the pressuriser water level is the CVCS flow balance—the balance between water being 'letdown' into the CVCS, and water being pumped back in via the charging flow and reactor coolant pump seals. (Have a look back at Chap. 7 if you need a reminder of the CVCS). On your PWR, there is an automatic control system that varies the charging flow to keep this all in balance (Fig. 16.2). If the pressuriser level is too low (for the current power level) the charging flow is increased until it's back where it belongs. If the level is too high, the charging flow is reduced to bring it back down. It's all quite slow, and you could choose to control this manually if you wanted.

The last thing you have direct control over in the primary circuit is its pressure. If you need to raise the pressure a little, increase the power to the pressuriser electrical heaters. This will boil-off a bit more of the water in the pressuriser, increasing the pressure in the steam bubble, and therefore the pressure in the whole of the primary circuit. If you want to lower pressure a little, just turn down the heater power. If you need to drop pressure a bit faster, open up the valves supplying the pressuriser water sprays. This spray water is taken from a cold leg, so is down at about 290 °C. It'll easily condense some of the 345 °C steam, dropping the pressure in the steam bubble. It's a lot faster and more dramatic than adjusting heater powers so use it with caution!

On your PWR, both the heaters and the spray valves can be driven under automatic control to hold primary circuit pressure at the desired value (Fig. 16.3). This will usually be a constant pressure of around 155 bar but may vary during heat-up and cooldown of the plant (see Chap. 21). You'll see in

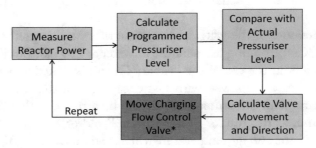

* NB: Charging Flow Control Valve OPENS if Pressuriser Level
is below programmed value, and CLOSES if it is above.

Fig. 16.2 Pressuriser level control system

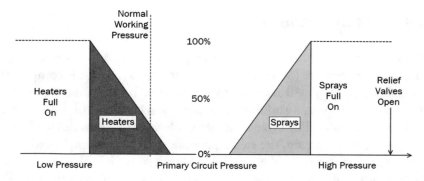

Fig. 16.3 Pressuriser pressure control system

Fig. 16.3 that a small amount of heating is required at normal working pressure. This offsets the trickle of water continually flowing through the sprays valve to protect them against thermal shock.

Finally, it's worth noting that on some PWRs there are venting valves or relief valves that the operator can open from the control room. If you have these on your plant, that'll be another way in which you can quickly reduce pressure in the primary circuit.

As an experienced operator there's a trick you can pull here if, say, you were making a change to the boron concentration in the primary circuit. If you're making a significant change there's a risk that the water in the pressuriser could get left behind at the 'old' boron concentration. This could then give you a surprise later-on, affecting reactivity if water comes out of the pressuriser after a trip.

Here's the trick: switch the heaters to manual control and turn them all on… Primary circuit pressure will rise as you put more steam into your bubble, but after just a little while the spray valves will begin to open. As the spray valves are still in automatic control, they will open just enough to balance the extra heating power, and the pressure will stop rising. Why bother to do this? Because you've now got much more water flowing into the pressuriser through the sprays, and out again at the bottom (to make-up for the water taken from the cold leg for the sprays); you are actively mixing the pressuriser water with the rest of the primary circuit, so it won't get left behind.

16.4   Steam Generators

Of course, there are other water levels for you to worry about—one for each of the steam generators (SGs). How these water levels change is complicated. At the simplest, they will respond to the balance between the water you're pumping in as feedwater, and the water that is leaving as steam. But, as you've seen, they also respond in a transient way to changes in steam demand. If you suddenly take more steam, the steam pressure and the pressure in each of the steam generators drops. As the pressure falls, the steam bubbles in the steam/water mixture will expand, causing the water levels to initially rise. Then they will start to fall as more water is being removed as steam, than is coming in as feedwater. There are other effects with SG water level changing in response to primary circuit or feedwater temperatures.

As I said, it's complicated, and no-one, as far as I know, operates a full-sized PWR with SG feedwater flows controlled manually. Your PWR has an automatic system controlling feedwater flows, and hence SG water levels, through feedwater valve positions and pump speeds (Fig. 16.4). It receives information from your power measuring instruments as well as temperatures and pressures, so it manages to control the SG water levels through all of the complications. Even so, SG water levels can change very rapidly in a transient or fault and will cause a reactor trip if they move to high or too low, outside of a very narrow acceptable range.

16.5   Steam Demand

Throughout this chapter on 'steady power', we've treated steam demand as fixed. In reality, it tends to move around a bit.

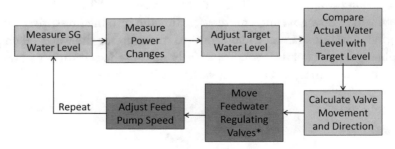

* NB: Valve selection depends on reactor power.

Fig. 16.4   Steam generator level control system (for one SG)

Your turbine will be operating under automatic control ('baseload'). All you have to do is set a desired output, in Megawatts, and the governor valves that admit steam to the turbine will be controlled to maintain a constant electrical output into the grid.

But, the grid isn't quite the fixed target that we'd like it to be. Grid frequency moves around. Not by much (usually by less than 0.1 Hz) but even those small changes will have an effect on your plant. As grid frequency changes, the speed of your turbine will change, as will the speed of all of your big pumps, including main seawater pumps, your feedwater pumps and even your reactor coolant pumps—each of these affecting the overall efficiency of the power station. Just to complicate things a little further, as the feedwater flow through the feed heaters varies, so does the amount of 'bled steam' that's needed to heat the water. This will affect the amount of steam left over, going through the turbine, to generate electricity, so that'll have another effect on efficiency!

In the control room, to maintain steady reactor power, you'll often find yourself having to adjust turbine output by a few MW either way. You'll notice this most often when the total demand on the grid is low (such as at night, when there's less connected equipment holding the frequency steady) or when demand is changing quickly. This is something that you have no control over, so you'll just have to get used to it.

16.6   What Else Might You Be Doing?

As you can see, you have many automatic control systems at your disposal, so you might think that you have lots of spare time? But you're running a power station with perhaps 200,000 items of equipment. Much of it is safety equipment that has no role in generating electricity, so will be on standby, rather than being in service. The only way to make sure that it will work when you need it is to test it. Much of your time will be spent in test-running pumps, valves, cooling equipment and protection systems. And of course you don't just test run it; you'll be measuring its performance, recording what you find and alerting the engineers to anything that looks like a problem.

There's a balance to be struck here. Too little testing and you'll have no confidence that the equipment will work when needed. On the other hand, too much testing can wear out equipment on its own and errors are sometimes made during testing leaving equipment unavailable without the operators' knowledge. Testing isn't without risk. You will probably find that the rules your station has agreed with your regulators will include some kind of testing schedule. So, as an operator, you won't have a choice but to do ithe testing on the agreed schedule or else shut down.

It doesn't stop at testing equipment. You have automatic systems that protect your plant and will shut it down if it moves outside its intended envelope. But, you don't want this to happen! So, you're going to be spending a lot of your time monitoring the plant and correcting anything that looks like it might be trending the wrong way. A good operator keeps the plant operating both safely and steadily, so avoiding transients and costly shutdowns.

Oh and things breakdown. There'll never be a day when something in that list of 200,000 plant items doesn't break. That's why you have two or four (or more) of each plant item—so, you can (nearly) always afford to have one breakdown, without it being a problem. Of course, much of this equipment will have planned maintenance routines, as well. You service your car regularly, to reduce the likelihood of it breaking down, and it's no different for the equipment at your power station. You don't just work with other operators; you need maintainers, and engineers, and planners and designers to replace obsolete equipment over the life of the station; physicists, chemists, radiation protection specialists, security guards and many others. There might only be a few people in the control room at any time, and it may not always be very busy, but there will be hundreds of people keeping your plant running safely.

Just stop and think for a moment about how much is involved in setting up the processes, procedures and training of personnel to make all of those activities happen in the right way. When people in the industry or outside it talk about a 'credible operator' they mean one that can do all of this. It's one of the reasons why countries often take many years to get a nuclear industry up and running, even when they've made a clear decision to go ahead with nuclear power.

16.7   Predicting Criticality

Imagine you're running your plant at full power… and it trips. It happens occasionally and, as you saw in Chap. 11, you need to know what to do next. In this instance, let's assume that the cause of the trip is found and corrected quite quickly, so now you're getting ready to restart your reactor. One of the things you saw earlier is that you'll be provided with a prediction, in terms of primary circuit boron, control rod withdrawal (and perhaps time) for when criticality would be reached. I could simply tell you that this is worked out by a computer, but a good operator can do the calculations for themselves!

If you've been monitoring your reactor closely, then just before the trip you'll know (or have recorded):

- The reactor power level
- The primary circuit boron concentration
- The control rod positions

All you then need to know is the time of the trip and the time at which you'll be ready to take the reactor critical again.

The thing you know for sure, though it's not in the list above, is that immediately before you tripped, the reactor *was* critical. In other words, all of the contributions to reactivity, before you tripped, must have balanced-out to give zero. If you can work out what's changed since just before the trip, you'll be half-way to predicting the next criticality.

So what's changed?

Firstly, you tripped! That means that all of the control rods fell in under gravity, so the contribution of the control rods to reactivity went from a small negative number (say, minus 20 milliNiles) to a very large negative number (perhaps minus 8000 milliNiles).

Next, you'll remember from Chap. 14 that there are reactivity changes associated with the temperature changes in moving up and down in power—the power defect. As you raise power, the power defect becomes progressively more negative as a reactivity contribution. If you trip, you get all of that reactivity back, as a positive contribution of around 1500 milliNiles (dependent on core burn-up due to changes in the value of the MTC).

The other significant change will be happening due to the post-trip xenon transient (Chap. 14), and that's what makes your criticality prediction time-dependent—at least to begin with. If you're not going to be ready to start-up for 3 days or more then you can assume that all of the pre-trip xenon will have decayed away. This will add around 2000 milliNiles of reactivity to your core and would make the calculation more straightforward, i.e. the xenon contribution would stop changing.

You'll likely be trying to start-up more quickly than this. If you take the reactor critical before around 20 h after the trip, xenon will be higher than before the trip so will give you a negative change in reactivity, after this it will be decaying away, and you'll have a positive change—in each case, it'll be varying with time, so you'll have a moving prediction!

You can see all of these effects in Fig. 16.5. On the left-hand side are the reactivity contributions immediately after the trip. On the right-hand side, the same contributions at the time you want to go critical—assuming here that we're past the 20-h point so that xenon is decaying away from its pre-trip value. For simplicity I've also assumed that the boron concentration is unchanged from the time of the trip.

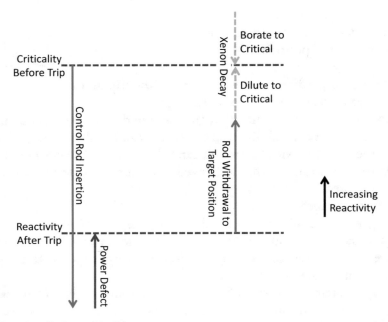

Fig. 16.5 Predicting criticality

With this diagram in front of you, you could simply pull the control rods until you achieve criticality. Unfortunately, this would probably take the control rods to a position that will adversely affect Axial Flux Distribution (AFD) as you raise power. From experience at your PWR, you'll be able to pick a control rod position for criticality that you know won't give you an AFD problem later on. This is your chosen point of criticality as far as rods are concerned.

Of course, a PWR gives you another way to control reactivity: boron. Look at the 'gap' on the figure with the green dashed line. I've shown it as a negative gap—meaning that your core is less reactive at the planned time of criticality than you want it to be. You can calculate how much of a change in primary circuit boron you need to make to bring the core back to zero reactivity (in this case a dilution).

If the gap were, say, minus 100 milliNiles, then that would be telling you that your core will be 100 milliNiles less reactive than you need. If you decrease the boron to give a change in reactivity of plus 100 milliNiles (about minus 15 ppm of boron), then the 'gap' will disappear, and you'll be able to hit your target. The 'gap' could be in the other direction: positive, shown by the orange dashed line. This would be telling you that you have too much reactivity and you'd have to borate to compensate. The nature of post-trip xenon transients is such that you could have a positive or a negative 'gap', but in either case, once you've performed the calculation, you adjust the boron, pull the rods and off you go!

Of course, if you don't have good records of the conditions before the trip, or the xenon was already in a complicated transient, then you may not be able to do the calculation by hand. You'd have to rely on the computer. In either case, as you've seen you'll always be approaching criticality cautiously, just in case the prediction is wrong.

17

It's All About Safety

17.1 The Interview

If you ever find yourself interviewing staff to join your power station, you might want to try this question:

"Is it Safe?"—Just that, nothing else.

Expect a pause while the interviewee struggles with how honest to be. Both 'Yes' and 'No' are unsatisfactory answers to the question. If they say 'Yes' it's either because they've failed to do any research, or because they think that's what you want to hear. If they think the answer's a simple 'No', then why on earth would they want the job?!

A better answer would be one that starts with "Yes, but…."

Nuclear power stations, including your PWR, are designed with safety as the paramount feature, but that does not mean they are entirely without risk. People who work at nuclear power stations are open and honest about both safety and risk. In this chapter and the ones that follow, you're going to see how important this behaviour is for someone who drives a reactor. But first I need to introduce the concept of a 'Safety Case'.

17.2 Building a Bridge

Imagine you've been given the job of designing a road-bridge. There are clearly a few design decisions to be made at the outset:

- How long/tall is it going to be? How many lanes of traffic will it carry?

© Springer Nature Switzerland AG 2019
C. Tucker, *How to Drive a Nuclear Reactor*, Springer Praxis Books,
https://doi.org/10.1007/978-3-030-33876-3_17

- What are you going to build it from—Steel? Concrete? Wood?
- What basic design will you use—Arch? Box Girder? Suspension? Cable-Stayed?

Now I need you to think of some practicalities:

- How robust will your bridge be? Are you designing it to cope with being filled with stationary cars? What about 40-tonne lorries? What about being filled with stationary lorries?
- What about abnormal loads? Could your bridge cope with a 150-tonne electrical transformer being carried across it, or will this need to find another route?
- How deep will your foundations need to be to support the bridge? What's the rock/soil like where you need to dig these foundations?

Then there's the environment to consider:

- Concrete and Steel behave differently as temperature changes, so what's the maximum temperature that your bridge structure is designed to cope with? What's the minimum temperature? What happens if the weather is worse than this?!?!
- Would the design of your bridge be affected by excessive amounts of rain or snow?
- How strong a wind are you going to design your bridge to withstand? Does its direction matter? Is your bridge going to be able to cope with a higher wind speed if only cars are permitted to cross?
- What does the bridge span? If it's a river or an estuary, could the foundations be eroded? Are tidal waves or storm surges a concern? If it crosses over a railway line, how would you cope with a derailed train striking the bridge?
- Now imagine a high wind, low temperature, abnormal load and a train derailment all happening at the same time...
- What about Earthquakes? How big? How frequent?
- Avalanche? Asteroid impact? Attack from the Death Star?

I'll stop there; I'm sure you get the idea.

17.3   Safety Cases

As you can see, before you start building the bridge, there are a whole lot of decisions to be made and justifications to be presented. You'll no doubt be writing this all down, and the suite of documents you produce will be your 'Safety Case' for the bridge (Fig. 17.1). The safety case argues why what you've designed is 'safe enough' in terms of materials used, construction methods, engineering standards etc. It'll cover what can go wrong and how your bridge will cope with it. A good safety case will also clearly show the boundary of what you are designing your bridge for; asteroid impact and 'Death Star' attack will probably be on the other side of this boundary so will be outside the 'Design Basis' for your bridge.

Bear in mind that there may be promises in your safety case—such as closing the bridge to lorries in high winds. That means you'll need to implement real-world restrictions; someone will need to monitor the wind speed and be able to close the bridge, or there's no point in writing it in your safety case! These promises on how you'll operate the bridge become a set of 'Operating Rules' which you must stay within. If you don't follow the rules, you're outside the safety case.

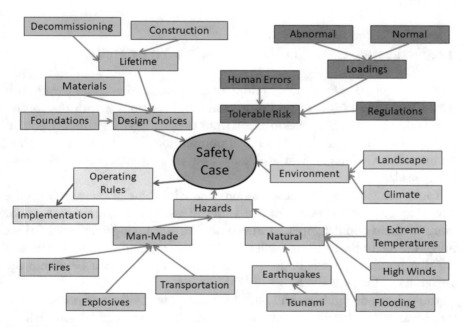

Fig. 17.1   Safety case

The idea of writing all of this down is that you'll then be able to present it to your regulators (planners etc.) when you apply to build. It's also a permanent record of what was agreed when the bridge was designed and helps future engineers to understand the path that you've chosen.

After considering all of that for a bridge, stop and think about how you'd do the same exercise when designing your PWR. For your PWR you'll have the same questions to answer on material selection, engineering standards and environmental hazards. But you've got a few differences to contend with. For a bridge, the 'risk' is the bridge falling down; for a nuclear power station, the 'risk' is an uncontrolled release of radioactive fission products to the environment (a 'radiological release'). In both instances, the key concern is avoiding harm to people, but there may be other concerns such as economic loss or damage to the environment. In your favour, there are more than 450 running nuclear power stations in the world and more being constructed. There are established 'Design Codes' for nuclear power stations, to which you can refer.

I've never seen a bridge's safety case, but I'd be surprised if it contained more than a couple of hundred documents. A nuclear power station's safety case will be held in tens of thousands of different documents, each covering just one aspect of the design. It typically takes 5–10 years, and hundreds of people, to design a nuclear station while developing its first safety case in parallel.

17.4   What Can Go Wrong with Your PWR?

You'll probably already be able to think of a few things that could wrong with your PWR (Fig. 17.2). It could shutdown (trip) spuriously, with the control rods falling in. It might develop a leak from the primary or secondary circuit. Perhaps the control rods could move unintentionally, or a reactor coolant pump fails? Or maybe you suddenly lose your connection to the electricity grid?

You're going to learn to recognise and deal with some of these faults in the next chapters, but for now, I'm just going to get you to assume that one or more of them could happen at any time. I'm not saying that any of these things are likely, though some will be more likely than others. I'm merely saying that none of them is impossible, so they'll probably each be within the 'design basis' for your PWR.

Your plant's design basis will have to encompass any fault or problem that is deemed by you (or your regulator) to be 'credible'. 'Incredible' faults will be those that are so unlikely that they can be discounted, provided that you can

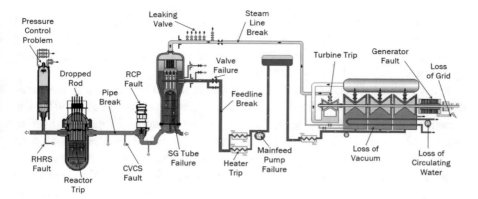

Fig. 17.2 Things that might go wrong?

also show that they can't be easily mitigated by changing your design. Even so, some 'incredible' fault combinations might still end up within your design basis—it's sometimes less work to include things than to argue that they can't happen!

17.5   The Three 'C's

There are just three things that you need to do to avoid the release of radioactive fission products to the environment. These three 'C's could be considered to be the pillars of nuclear safety:

* Criticality; strictly speaking, it's sub-criticality, i.e. shut down the reactor!
* Cooling; remove the decay heat from the core to stop it being damaged.
* Containment; maintaining the integrity of the fuel (clad and fuel pellets), the primary circuit and the reactor building. These three form the three barriers to fission product release. Any one of these, on its own, would be sufficient to prevent a release, so protecting them is ultimately the key to the safety of the people who live around the station.

17.6   Automatic Protection

Do you think people are serious about speed limits in your country?

I'm not sure that they are. Our cars and other vehicles rarely include devices that control our speed to keep us within limits (I'm ignoring 'Cruise Control' as its use is optional). What if I suggested fitting something to your car that would automatically shut down the engine if you went above the speed limit? After grinding slowly to a halt, you'd have to start it and pull away all over again; presumably a bit more carefully.

Perhaps there are good reasons for not doing this in cars (it could cause collisions?), but this is the mental model you need when driving your PWR. The system that does this for you is known as the 'Reactor Protection System (RPS)'. On your PWR the RPS continuously monitors the status of your reactor and connected equipment. If the signals that it is monitoring move outside of predetermined limits ('setpoints') it will shut down the reactor. As an operator in the main control room, *you cannot override this*. It is a built-in feature of the plant and will be relied upon heavily in your safety case. So, the only way you have of avoiding an automatic shutdown (trip), is to drive your reactor within limits! That might sound difficult but, with practice, it's not unusual for a modern PWR to run for 5–10 years without an automatic trip.

You'll see in Fig. 17.3 how the RPS monitors a lot of plant parameters. The signals from these instruments feed into electronics that compares them with the chosen setpoints. If any of the parameters move beyond a setpoint for a reactor trip, circuit breakers are opened in the path from the power supplies to the control rod drives. The control rods fall into the core, and you have an automatic trip!

In reality, you're going to have multiple instruments for each parameter. Four is typical. Each will feed into a different 'Guardline', and then a further bit of electronics will combine the outputs in a voting system. If the voting is arranged as 2-out-of-four then a trip will occur if any two out of a group of

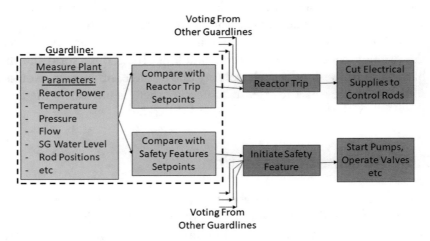

Fig. 17.3 Reactor protection system (RPS)

four instruments show that a setpoint has been reached. Why design it this way? It allows for instrument unreliability—you wouldn't want a reactor trip to occur spuriously because a single instrument has failed to a state that looks like it's over the setpoint.

It also means that you can take one instrument out of service for maintenance or testing but still be well protected by the availability of the other three (now operating on a one-out-of-three, or two-out-of-three basis depending on your specific electronics design). It means a lot more instruments to buy, maintain and replace when they wear out, but you'll see enormous benefits in plant reliability over the station's life if you do it this way, and the risk of your RPS failing to 'see' a fault, if one occurs, will be very much reduced.

17.7   Engineered Safety Features

Have another look at Fig. 17.3. You'll see that there's another set of outputs—apart from the reactor trip signal. These go to the 'Engineered Safety Features (ESFs)'. Faults at PWRs can develop in reasonably short timescales—just a few minutes—so even if the plant trips, it wouldn't be fair (or even possible) to rely on you (the operator) to perform all the necessary plant actions to ensure that the three 'C's are met. Instead, on PWRs, the RPS has been expanded to fulfil other functions.

Let me give you an example: the level of water in each of your Steam Generators (SGs) is important for safety. Too high and you risk water being carried over to your turbine—which could destroy it, violently. Too low and you've lost the ability to remove heat from the primary circuit. Naturally enough there's a small acceptable range for SG water level, and if any of the four SGs move outside of this range the RPS will initiate an automatic trip. Then what? You still have a high or low water level, so something needs to be done about it…

Your PWR will be provided with a lot of safety equipment. This will include back-up (or 'Auxiliary') feedwater pumps for use when the SGs aren't getting enough feedwater from the main feedwater pumps. So, below or close to the reactor trip setpoint on low SG water level, there'll be another setpoint that initiates the 'Auxiliary Feedwater System' pumps if the level does not recover after the trip. In the other direction, you'd expect to see more setpoints that restrict or isolate feedwater supplies to SGs if the water level continues to go high after a trip. The point here is that these additional setpoints actually have nothing to do with the trip setpoint—they'll still cause actuation of auxiliary feed, or restrict/isolate feedwater if that is required by the SG levels, even if the plant tripped days or weeks ago!

On most PWRs, you can look at the ESFs (the safety systems initiated by the RPS or by the operator) and divide them into three groups:

- Emergency Core Cooling Systems (ECCS).
- Containment Systems.
- Others (including, for example, the Auxiliary Feedwater Systems).

You'll meet ESFs in each of these categories when you look at faults in more detail in the next chapters, so I won't describe them here. The one ESF signal that I will mention now is something called a 'Safety Injection (SI)'. This is best described as the 'Seventh Cavalry' of the ESFs because in reality it's a whole bunch of safety system initiations rolled into one. You won't always see an SI signal when you trip—it's actually quite rare—but you'll always trip if you do get an SI. For any of the more significant faults, an SI signal will do everything you need to meet the three 'Cs', at least to begin with.

17.8   How Safe Is 'Safe Enough'?

In any industry, this is a difficult question, but we can't avoid trying to answer it. As a designer you can keep adding layers of safety; better, stronger containments, more backup-systems, more instruments, more pumps, valves etc. But this makes the plant more and more expensive. It also has diminishing returns as far as safety goes. If there's too much equipment, it becomes so complicated that errors will be made or there will be combinations of failures that you hadn't planned for. Having one of something (say an 'Emergency Core Cooling System (ECCS)' pump) is a start. Two is much better as it allows for one to fail. Three is excellent as one can fail, one can be out of service on maintenance, and you still have one leftover. Four of something is even better, but not a lot better. Adding a fifth pump would make very little difference to risk.

Most modern plants have safety systems that are quadruplicated, i.e. four sets of electronics, four sets of pumps, valves, pipework, four sets of power supplies etc.—in other words, a 'four safety train' plant. No-one as far as I'm aware is planning a plant with five-fold safety systems. Looked at another way, there seems to be a broad consensus that 'four-train' plants are safe enough… but who actually made that decision for your PWR? Let's look at a bit of history.

17.9   The Windscale Fire

When early reactors were built in the UK, they were considered a military project and did not benefit from strong regulation. The 1957 fire at Windscale Pile No.1 changed that.

The Windscale Piles were natural uranium fuelled, graphite-moderated, air-cooled reactors producing plutonium for nuclear weapons. Imagine, if you will, a stack of 2000 tonnes of graphite blocks, 7 m high and 15 m in diameter; a lot bigger than CP-1 (Chap. 4). Each 'Pile' (there were two) accommodated 3440 horizontal channels into which natural uranium metal fuel, encased in thin aluminium cans could be pushed. The principle of the design was that air would be blown through the channels to remove heat while the reactor was running. Some of the uranium-238 would be transformed into plutonium-239 and the irradiated fuel, pushed out of the back of the reactor, would be transported for reprocessing to extract the Plutonium (Fig. 17.4).

Unfortunately, if you bombard graphite with neutrons, some of the carbon atoms in the graphite will be knocked out of their regular positions, storing 'potential energy'—energy that can be released if they fall back to where they belong. This stored energy is known as 'Wigner Energy' after the physicist who discovered it. If you irradiate graphite at a high enough temperature, such as in a MAGNOX or AGR, (see Chap. 22) then this is not much of a problem, but the Windscale Piles usually ran at low temperature, so regular, deliberate temperature increases were necessary to allow the Wigner energy to dissipate.

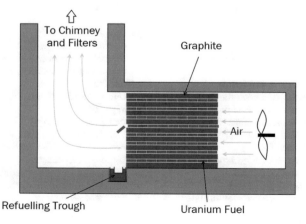

Fig. 17.4   Layout of a windscale pile

It was during one of these temperature increases that things got out of hand. More Wigner energy was released than was expected, and the fuel and graphite ignited, setting a large part of Pile 1 on fire and releasing significant quantities of fission products into the cooling airstream. This was partially filtered, but even so, there was a significant uncontrolled radiological release, with full dismantling of the Piles taking many decades.

The Windscale fire was a bit of a wake-up moment for the UK and led to the first (1959) Nuclear Installations Act. This required that the civil nuclear power stations which were then under construction (e.g. the MAGNOX plants) and those planned for the future, should be licensed by the newly formed Nuclear Installations Inspectorate (NII); a regulator whose sole responsibility was Safety. The NII's functions are today carried out by the Office for Nuclear Regulation (ONR)—I guess in modern parlance that would be 'OffNuke'?

Why does this matter to you? Because it is the ONR that will have decided that your PWR design is 'safe enough' to be built and operated. In the UK, nuclear power stations (and other nuclear installations) are given permission to be constructed, operated and dismantled ('decommissioned') through a licencing processing managed by the ONR. Your PWR will have its own 'Site Licence' setting out the boundaries of your site and the design of the reactor that you're driving. The site licence will include a list of conditions (Fig. 17.5)

1. Interpretation
2. Marking of the site boundary
3. Control of property transactions
4. Restrictions on nuclear matter on the site
5. Consignment of nuclear matter
6. Documents, records, authorities and certificates
7. Incidents on the site
8. Warning notices
9. Instructions to persons on the site
10. Training
11. Emergency arrangements
12. Duly authorised and other suitably qualified and experienced persons
13. Nuclear safety committee
14. Safety documentation
15. Periodic review
16. Site plans, designs and specifications
17. Management systems
18. Radiological protection
19. Construction or installation of new plant
20. Modification to design of plant under construction
21. Commissioning
22. Modification or experiment on existing plant
23. Operating rules
24. Operating instructions
25. Operational records
26. Control and supervision of operations
27. Safety mechanisms, devices and circuits
28. Examination, inspection, maintenance and testing
29. Duty to carry out tests, inspections and examinations
30. Periodic shutdown
31. Shutdown of specified operations
32. Accumulation of radioactive waste
33. Disposal of radioactive waste
34. Leakage and escape of radioactive material and radioactive waste
35. Decommissioning
36. Organisational capability

Fig. 17.5 UK ONR site licence conditions (2019)

that your company has to meet. These include having a safety case, training of the operators, maintenance of the plant, emergency planning in case there is a release of radioactivity offsite, operating rules (taken from the safety case, as for the bridge design, above), a formal process for managing modifications to the plant, and so on.

These conditions have the force of law behind them, so as a Licence holder, your company must comply with them. If you breach them, the Station Director can end-up in court. He or she will not be amused.

17.10   International Perspectives

Other countries had different paths to regulation, some following their own events, others by learning from their neighbours' experiences. Their regulators will work in subtly different ways, and their licencing systems will differ, but they all have similar underlying principles. Two organisations help with this, the first is the International Atomic Energy Agency (IAEA), a body set-up by the United Nations to promote the safe, secure and peaceful use of nuclear technology. The IAEA publishes Standards and Guides, enabling both operating companies and regulators to easily make international comparisons in what they do.

The IAEA also provides a 'Safeguards' inspection function under the Nuclear Non-Proliferation Treaty (see Chap. 23), though the European Union offers its own inspection regime through the EURATOM treaty (note: EURATOM is *not* a nuclear safety regulator; it primarily exists to keep track of nuclear fuel).

The other organisation that provides an international view on nuclear energy is the World Association of Nuclear Operators (WANO). WANO was set-up in the wake of the Chernobyl disaster (Chap. 9), and its approach is focused on 'best practices' collected from operating companies around the world. When WANO visit your site- and they will—they will tell you what your current 'Gaps to Excellence' are, and discuss with you how you plan to address them. If you're currently working in an industry other than the nuclear industry, you might want to stop and think about that. How would you feel about paying for a couple of dozen of your international competitors to visit your plant, examine it (and you) in detail and tell you what you should be doing better? That's normal business at a nuclear power station.

Back in the UK, the ONR is the regulator who will visit your plant, inspect it and oversee any significant projects or changes to the design. They, like their counterparts in other countries, are powerful regulators. If you step out of line and are found to be operating your plant poorly or operating outside of your safety case, they can (and will) shut you down. But their knowledge and experience can be useful to you, so if you meet one of their inspectors while driving your PWR, be open and honest with them, hide nothing, and listen to their concerns and advice. You will benefit from their point of view, and they will be reassured if they can see that you have been well-trained in your role.

17.11   Tolerable Risk

Broadly speaking, the risk from a plant is deemed to be 'Tolerable' if it can be shown to be very small compared with the dangers that people are exposed to on a day-to-day basis (e.g. from driving, eating, working etc.), *and* to reduce the risk further would be grossly disproportionate in terms of time or cost, compared to the risk-benefit. In (very) round numbers: the risk of death to an individual might be tolerable if it's around one in a million per reactor per year. But this is a moving target, so don't think that you've finished the job, just because you've got your site licence. This is why, in general, newer plants are usually deemed to present a lower risk to the public than older ones. The technologies have improved, and the targets have got tougher.

Throughout the life of your station, you're going to have to be able to demonstrate to your regulator, that you're striving to improve your safety, and that you're responding to events and developments in knowledge, i.e. that you're trying to reduce risk to a level that's 'As Low as Reasonably Practicable (ALARP)'.

That's what's meant by 'Safe Enough'.

17.12   Just a Small One…

A few years ago, I received a phone call from a TV researcher. I won't say who he was working for, but the idea he outlined was this: to find a village somewhere in the UK, get the residents to build a small nuclear reactor (with our help) and generate some electricity to supply the village. The idea was great; de-mystify nuclear energy and show people how safe it can be. It would probably have been lots of fun as well!

However, as I explained, the project was going to hit three major problems—all of which you should by now appreciate:

Firstly, it's very challenging to build a small reactor. Leakage of neutrons out of the sides means that there's a practical lower limit to the size of any reactor if you want to achieve criticality. The only way to make a really small reactor is to use highly enriched uranium or plutonium fuel—and that's not going to be available to the villagers, as you'll see in a later chapter.

This second problem is engineering. Reactors only generate electricity because we use them to produce steam; and that in turn needs high temperatures and pressures. While you might (perhaps) trust a group of supervised villagers to build a boiler for a steam engine, would you trust them to construct a pressure vessel to contain a nuclear reactor, including the radioactive fission products?

Finally—and this was the killer—there's Licencing. There is no get-out for a 'small' reactor in the UK Nuclear Installations Acts. The only way to build a reactor is to first satisfy the ONR that you can meet all of the site licence conditions that they'll impose. It takes years of work and millions of pounds to reach that point, to become a 'Credible Operator'. That is not something that any group of amateurs, however keen, will ever accomplish.

The researcher was, unfortunately, disappointed.

18

What Can Go Wrong
(and What You Can Do About It)

18.1 Can You Cope?

At full power, your reactor produces 3500 Megawatts (MW) of heat. Your power station successfully turns some of this heat into electricity. As a trainee reactor operator, you've learnt how to start up the reactor and move power around. As you've seen, there will be a few other tasks for you to perform in the control room, but at this point, you might be thinking how easy it all looks. Why would it take so long to train a real reactor operator?

Learning to drive a reactor is a bit like learning to fly a plane. Most people get the hang of take-offs and landings pretty quickly. What takes the time and effort, especially for large passenger aircraft, is learning how to respond when something goes wrong. It's the same with a reactor. Your plant has many automatic systems that help you to drive it. It also has monitoring systems that you can rely on to shut down the reactor and perform other protective actions in the event of something serious happening. But, as an operator, you will have an essential role in controlling the plant, avoiding problems (if they can be avoided), responding to events and minimising their possible effects on the environment. This is why so much of your training is taken up with abnormal operations or 'faults'.

What happens if you get it wrong? Your PWR can't explode—you're not trying to drive a nuclear bomb—and in many ways, it's a very stable machine. But it's still possible for you to take actions—especially during a fault—that might adversely affect the cooling of the reactor. The decay heat of a recently shutdown reactor is in the tens of MW, so if you fail to cool it, the nuclear fuel will be damaged and release radioactivity into the primary circuit. From there it might find its way into or even out of the reactor building: a very bad day in the control room.

The three 'C's that you met in the last chapter are the keys to reactor safety—(sub)Criticality, Cooling and Containment. Bear these in mind as you learn about each of the faults in this and the next few chapters, and you'll be well on the way to understanding the reactor operator's role.

The final thing to remember before we start looking at some examples of possible faults is that that's all they are: 'possible'. Possible doesn't mean 'likely'. There are some events that are likely to happen over the decades of life of your nuclear power station. Others have never happened to any nuclear power station—and probably never will. In between, there is a range of faults that have occurred occasionally to stations worldwide. Generally speaking, the more severe a fault is, the less likely it is to happen; or in safety case terms, 'less frequent'. I'll try to give you an idea of likelihood (frequency) as I explain each of the faults, but, as an operator, it doesn't really make a difference to you. You have to learn to deal with both the likely and unlikely events.

18.2   Fault 1: A Loss of Grid

Let's start with something easy. Imagine that you're driving your reactor at steady, full power when, without any warning, you lose connection to the electricity grid. Why might this happen? Perhaps a storm has damaged some grid lines, or a large substation has suffered a fire? If these problems affect enough gridlines on your area, you might suddenly find that you're not connected to the grid any more.

How likely is a loss of grid? Well, it varies tremendously from plant to plant. Some go their entire lives never losing offsite connection, others seem to average an event every couple of years. The two important things in assessing your plant's loss of grid frequency will be the overall reliability of your country's grid system, and the length of the electricity lines to which your station is connected. Why? Because there's a strong link between line-length and the likelihood of an electrical fault, with longer lines seemingly more susceptible. By the time you've been running your plant for a couple of decades you'll probably have a realistic view of your local loss of grid frequency.

The first things you'll probably see in the control room are the alarms and indications of a turbine and reactor trip. How you respond to this was covered in the chapter on the 'Big Red Button', but you'll pretty quickly see that this isn't just any old reactor trip… If you've lost the grid and your turbine is tripped, then there will be nothing to supply the High Voltage (HV) electricity needed for the motors of all of your large pumps. Depending on your plant design, I'm going to guess that you'll have lost electrical power to:

- The circulating water pumps that supply sea-water to your turbine condensers
- The main feedwater pumps that usually supply the steam generators (SGs)
- The reactor coolant pumps (RCPs)

Losing the sea-water pumps means that vacuum will degrade very quickly in your turbine condensers. For the first few minutes after the trip, you might be able to dump steam to these, but once enough vacuum is lost, that won't be possible, and you'll have no option but to dump steam to atmosphere through your power operated relief valves (PORVs), mounted on the main steam lines.

Losing the main feedwater pumps sounds serious, but it's not too much of a problem. The reactor trip has stopped the chain reaction, so the only heat you need to remove from the core is decay heat; many MW, admittedly, but still only a few per cent of full power. Your smaller auxiliary feedwater pumps can supply enough feedwater to the SGs to manage this—and we'll talk about how these pumps are powered a little later.

What about the reactor coolant pumps? The RCPs have large electrical motors; it's not feasible to keep these running when you lose the grid. Each RCP has a flywheel so won't stop immediately, but will run-down over just a few minutes. Aren't we relying on these to drive water through the core and cool the reactor, even when it's shutdown? Well, yes, usually, but on a loss of grid, that's not an option. Instead you're going to have to trust to physics.

18.3   Natural Circulation

As the RCPs slow down and stop, the flow of water through the core reduces. As it does so, the amount of decay heat transferred to each kilogram of water will rise, so the temperature difference (Thot minus Tcold) will increase— water at the top of the core in your tripped reactor will be hotter than it would have been with the RCPs still running. Warmer water is less dense than the colder water in the SG tubes so the colder water will tend to fall out of the tubes displacing the warmer water. The geometry of the primary circuit is ideal for this as you have a hot reactor low down and colder SGs higher-up. Even more helpfully, there's an unobstructed flow path from the top of the reactor, through the hot legs, to the SGs tubes, then from the SG tubes back along the crossover and cold legs to the bottom the reactor. We call this flow of water around the primary circuit 'natural circulation' as it is driven only by temperature and density differences rather than by pumping (a bit like the secondary side of the SGs).

As the temperature difference across the core rises, so does the natural circulation flow, as the density changes become greater. Eventually, the natural circulation will be large enough to remove all of the heat from the reactor, and the temperatures will stop rising. Typically your PWR will reach this point in less than 15 min after a loss of grid, with a temperature difference of about half of that you see at full power, as you can see in Fig. 18.1.

You don't have to control any of this; the physics means that the primary circuit will reach its own equilibrium. Even better, as the decay heat falls, the natural circulation will fall along with it, again without any control operations on your part. You can see this starting to happen in Fig. 18.1, as Thot is just beginning to fall.

You might be wondering what keeps Tcold steady throughout this event. It's steam dumping via the PORVs. The steam dumping holds steam pressure in the SGs constant. This fixes the SG temperature as the SG is stuck on the boiling curve, and this, in turn, fixes Tcold.

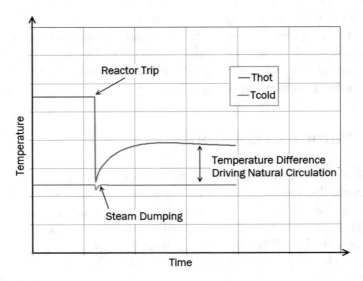

Fig. 18.1 Natural circulation

18.4   Batteries and Back-Up Generators

It's worth thinking about back-up power. In your home, you probably have very little in the way of back-up power, though perhaps some of your electronic devices have internal batteries, so will keep working (or at least not forget the time) if you have a power cut. You probably don't have a back-up generator, unless you live somewhere especially rural.

The need for back-up power is quite different at a nuclear power station. You have lots of instrumentation, computers, controls and protection systems which you don't want to lose on a loss of grid, so that means that you also need to have some large batteries ready to provide uninterrupted low voltage supplies for all of this equipment. But batteries only last so long, and you're probably going to want some higher voltage supplies to run pumps, e.g. auxiliary feedwater pumps and CVCS pumps for RCP seal injection and boration. In other words, you need some large back-up generators. Some plants use gas turbines (jet engines) for this, but diesel generators are more common for PWRs worldwide.

Your PWR is typical. You have four large 'Essential Diesel Generators (EDGs)', each capable of producing between 5 and 10 MW of electrical power (an example is shown in Fig. 18.2). That probably seems a lot, but as

Fig. 18.2   Essential diesel and generator

you'll see, for a more significant fault, they have to power a lot of equipment. (Though it's still not enough to start an RCP.) Why have four EDGs? It's a similar approach to other safety systems. Your safety case should be able to show that you can cope with any fault (including a loss of grid) with just one or two working EDGs, so having four to start with enables you to take one out of service for maintenance, and still have one more EDG fail to start, without it being a problem.

18.5 Pumps etc.

Once your EDGs have started—and you'd expect that to happen very quickly and automatically when the grid is lost—you'll have electrical power available to keep your batteries charged and to run some of the medium-sized pumps such as auxiliary feedwater pumps and CVCS pumps. You can't switch all of this equipment on at the same time though as that would stall the diesel. Instead, those automatic systems will have to sequence the electrical loads onto the engine over a few minutes, so that the diesel can respond without stalling.

There's another way of providing power to pumps: steam. You have lots of steam, and at the moment, it's all being dumped to atmosphere. Instead, you can divert some of the steam to small turbines driving pumps. Your PWR has a few of these, some as a diverse design of auxiliary feedwater pumps, others as a diverse source of injection water for the RCP seals. Importantly, these steam-driven pumps require no EDG supplies to operate, so would be available to you even if none of your EDGs started on a loss of grid. It's worth saying that some reactor designers don't like steam-driven pumps as they tend to need a fair bit of maintenance. Instead, to provide a bit of diversity, these designers might give a reactor more diesel generators, some being built to different designs to ensure that no problem can affect all of your diesels at the same time.

18.6 Recovering from a Loss of Grid

You can't start-up your PWR without a supply of electricity from the grid. Just trying to start an RCP would stall your EDGs—even if you could connect them. You need all four RCPs running before your reactor protection system (RPS) will let you reset the trip, so you have no 'black-start' capability, unlike some coal and gas stations that can start-up without grid supplies.

You're not going to be able to help to restart the grid in your area; you'll have to wait until it is returned to you by others. This is a good argument for having lots of fuel oil (for the EDGs) and water (for the auxiliary feedwater pumps).

Once the grid is restored, you might think you could just re-connect your station's electrical system to the offsite supply. In reality, this would cause a very similar problem to trying to connect all of your loads to a running EDG. This time, rather than stalling an engine, you'd find that the current would be too high from trying to start-up all of the equipment on your system at once—you'd blow a fuse (or some other more sophisticated kind of electrical protection). Your only option is to start at the highest voltage electrical boards on your station and disconnect everything leading from them. When you've done this, you can safely connect those boards to the grid, then reconnect the out-going circuits one at a time. You'll need to repeat this every time you try to re-energise a board, all the way down to the lowest voltage boards and equipment.

It won't be quick—it'll probably be 24 h before everything is up and running again. Then you can think about starting-up your reactor.

18.7   Fault 2: A Large Break Loss of Coolant Accident (LB LOCA)

A Large Break Loss of Coolant Accident is a peculiar fault. No-one has ever had one, yet it's the text-book fault which every PWR is designed to cope with. In some ways, it's the worst fault that can happen within the design-basis of your plant, so you can take comfort from the fact that if your design can cope with a Large LOCA, you can be sure that it can deal with (almost) everything else that isn't as large… Bear that in mind as you think about the text that follows. Incidentally, I find that LOCA is usually pronounced as 'Low-ker', but you sometimes also hear it as 'Lock-ah'.

What do I mean by a Large LOCA? I mean a break in the primary circuit up to the size of a complete fracture of a hot or cold Leg. That's a big break. It's going to be messy.

A substantial breach in the primary circuit will cause primary circuit pressure to fall very quickly. It'll cause a loss of moderator between the fuel pins as the water rapidly boils. You won't be cooling the fuel so effectively, so fuel temperatures will rise. Your PWR has strongly negative fuel temperature and void coefficients (Chap. 9), so this will give lots of negative reactivity—power will drop very fast even before the RPS sees the problem and sends out a trip signal, dropping in the control rods.

The good news is that your reactor is now shutdown. The bad news is that the core will almost certainly have dried-out, so the fuel won't be being adequately cooled. You need to get water into the core to re-cover the fuel and start removing heat. How long do you have in which to do this? A few minutes. After that, fuel would begin to be damaged, and you'd be releasing fission products into the reactor building.

No-one expects operators to diagnose a fault and respond by actuating the correct safety systems within just a few minutes—30 min is a more typical safety case claim. This means that your RPS will have to do the job for you. It will see the sudden drop in primary circuit pressure and the sudden *rise* in reactor building pressure (as primary circuit water turns to steam) and it will very quickly decide that both a reactor trip and a 'Safety Injection (SI)' signal are required.

18.8   Safety Injection (SI)

The most important function of an SI signal is to do what the name suggests—inject water into the primary circuit. It does this by starting all of the 'Emergency Core Cooling System (ECCS)' pumps. Again, with more pumps, there's a good chance that enough of them will be working, even allowing for maintenance and breakdowns.

If you're a mechanical engineer, you probably already know about 'pump curves'. For the rest of us, here's a simple explanation. Most large pumps operate by spinning one or more propeller-like objects (impellors) in an enclosed volume such as a pipe or pump-bowl. The water in the pipe or bowl is flung outwards by the impellor and builds-up pressure and speed in the process. Such a pump is known as a 'Centrifugal Pump'. Centrifugal pumps always have a 'pressure versus flow' characteristic something like the ones in Fig. 18.3—for two differently designed pumps. For each pump, the higher the pressure that the pump is delivering against, the lower will be its flow. Conversely, if the pressure it's pushing against is very low, then you'll get maximum flow. This is an important concept when it comes to dealing with Loss of Coolant Accidents (LOCAs) as you'll see in the next chapter.

An analogy—and this is not an exact analogue, but just for illustration—turn on a garden hose to full flow. The water is coming out quickly but doesn't feel like it's at very high pressure. Now put your thumb over the end of the hose. The chances are that you won't stop the water flow entirely, some will

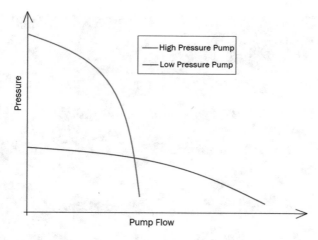

Fig. 18.3   Centrifugal pump curves

still force its way out. It'll be a much lower flow, but it'll be coming out at much higher pressure than before the thumb was applied. Oh, and you're probably now very wet, sorry.

It's hard to find a single pump that can cope with all of the possible faults on a PWR, where the primary circuit pressure could be higher than 100 bar for some faults, or almost nothing in a Large LOCA. More usually, ECCS pumps are designed in two (or more) groups, e.g.:

- 'Low Pressure Safety Injection' pumps designed to give a high flow but unable to deliver at high pressure, and,
- 'High Pressure Safety Injection' pumps giving a lower flow of water but able to deliver at very high pressure.

Figure 18.3 shows how the pumps curves will differ for a high pressure and a low pressure pump. Figure 18.4 shows an example of a real low pressure safety injection pump. This is an electrically driven pump, with a large (500 kW) motor.

Your SI signal starts both sets of pumps, so that water can be delivered to the primary circuit, re-flooding the core and restoring cooling, regardless of how much pressure is left in the primary circuit, or how it is changing.

Even so, electrically driven pumps take a little while (seconds) to get up to speed—longer if you were unlucky enough to lose the grid and have to wait for your EDGs to start-up before the pumps can be started. Because of this

Fig. 18.4 Low pressure safety injection pump

possible delay, your ECCS also includes four 'Safety Injection Accumulators' inside your reactor building.

Each accumulator is a simple tank of borated water, held at a pressure of around 40 bar by nitrogen gas in the top third of the tank. Between the accumulators and the primary circuit there are just a couple of non-return valves (flaps that only allow flow in one direction). These valves are held firmly shut all the time that primary circuit pressure is higher than accumulator pressure. However, if primary circuit pressure falls below 40 bar (such as in a Large LOCA), the nitrogen pressure in the accumulators forces open the non-return valves, and the water is pushed into the primary circuit. No electrical power and no pumps needed, just physics!

Figure 18.5 is a sketch of part of your ECCS, as connected to one of your cooling loops. The pumps initially deliver water to the cold legs, though this can be switched to the hot legs later if desired. They take their water from a large tank outside of the reactor building. This is the same borated water you'd use when you refuel the reactor (Chap. 21), known as the 'Refuelling Water Storage Tank (RWST)'.

The safety injection accumulators won't re-flood the core on their own but will have given you a good head-start before the safety injection pumps come online. So, by the end of the first couple of minutes, you'd expect your core to

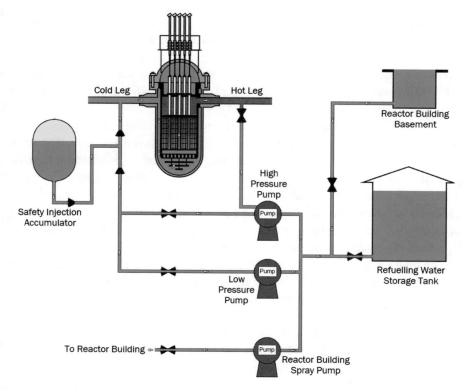

Fig. 18.5 Emergency core cooling system

be re-flooded and water to being pumped into the primary circuit. Of course, it's still pouring out of the break and into the reactor building, but never mind.

Although it's not strictly part of the ECCS, you have some other pumps that give you a 'Reactor Building Spray system'. The RPS will start these pumps in response to a significant rise in reactor building pressure. The spray pumps deliver RWST water to spray nozzles mounted on the inside of the reactor building roof. This spray of cold water helps to condense the steam inside the building and stops the pressure from getting too high. At some plants the same function is provided by water flowing down the outside of the reactor building.

Stop and think about how much you, as the operator, have had to do in these first few minutes of the Large LOCA…. actually, nothing. All of this happens so quickly that you can't be expected to do anything. For a significant fault, it's typical for the operator to be playing catch-up for a while, while monitoring and assessing their indications.

A Large LOCA challenges equipment and instruments inside the reactor building as they will suddenly be exposed to high temperature, pressure, humidity and radiation levels. Anything you need to use after a Large LOCA needs to be 'qualified' (e.g. tested), to cope with these conditions. As an operator, you'll need to know what is qualified and what isn't, and not try to use anything in the second category.

If your plant has been well designed and you're using good quality fuel, you'll find that your core will have come through even a Large LOCA with very little damage. It's a significant transient, involving rapid changes in temperature and pressure, but it's what the fuel clad is designed to cope with. There may be a small number of failed fuel pins that release some fission products as gases into the primary circuit, and from there into the reactor building, but you'll remember that one of the functions of the SI Signal was to isolate the reactor building to stop any leakage out into the environment.

The final piece of the Large LOCA jigsaw comes when the RWST gets close to being empty. You'll still be losing water out of the break, so if you just stop pumping water into the primary circuit, the core will dry out all over again. But wait a minute… Where has all that water gone? It's ended up in the basement of the reactor building. So, as the level of water in the RWST falls, it is monitored by your RPS. When the level is low enough, the RPS operates some valves, and your pumps start taking water from the basement instead! The water is run through some heat exchangers to take away the heat, but otherwise, it's simply recirculated through the primary circuit, out the break, and back into the basement. Over the coming months, as decay heat falls, you'll gradually be able to reduce the number of running pumps, and eventually be able to get into the reactor building to clean it all up.

In some of the more modern PWRs, the basement of the Reactor Building *is* the RWST, so there's no changeover required to achieve recirculation, but the layout I've described is currently more common.

It's worth noting that in all of this, even if there's been a little bit of fuel damage, there's been no release of radioactivity to the environment. The reactor building is a very effective barrier to release, even in a significant fault. Honestly though, if yours were the first plant in the world to suffer a Large LOCA, that would suggest that there's something very wrong with the way it was designed, maintained or operated. Don't expect to ever start it up again.

19

Smaller Isn't Always Easier

This might surprise you: A Small Loss of Coolant Accident (Small LOCA) might be harder for you to manage than a Large LOCA.

19.1 Fault 3: A Small Loss of Coolant Accident

Your primary circuit has lots of connections. For example, there are the CVCS charging and letdown lines and the pipework for the emergency core cooling system. There are also a large number of small connections for chemical sampling and for the instruments that measure the primary circuit's temperature, pressure and flow. All of these connections have been carefully constructed and will be regularly inspected, but it's just possible that one of them could fail and start leaking. What then?

This won't be a fault that looks like a Large LOCA. You're not going to see the core empty of water in just a few seconds with the primary circuit pressure falling away to almost nothing. Instead, you'll see a range of symptoms that'll give you a clue to just how small the Small LOCA might be.

For the smallest of LOCAs, you won't see the primary circuit pressure fall at all. Your first indication could be an unexplained drop in the water level in your CVCS volume control tank (VCT). Is that surprising? Well, imagine that the leak is so small that the only effect it has (to begin with) is to give a slight drop in pressuriser water level. The pressuriser level control system will see this fall and compensate by opening up your charging flow control valve— see Chap. 7 for a reminder of the CVCS—that is, it will increase charging flow to hold pressuriser level steady. But now, your CVCS is charging more

© Springer Nature Switzerland AG 2019
C. Tucker, *How to Drive a Nuclear Reactor*, Springer Praxis Books,
https://doi.org/10.1007/978-3-030-33876-3_19

than it's letdown flow, so the VCT level will fall. To be fair, you'd also expect to see some alarms from the reactor building reporting higher humidity and radiation levels, but these indications would be typical for a leak of a few litres per minute from the primary circuit—what you might call a Very Small LOCA.

If the Small LOCA is just a little bit bigger, your CVCS charging flow won't be able to be increased high enough to compensate for the loss of primary circuit water. This means that the pressuriser water level is going to continue to fall. As the water level drops the steam bubble above the water will expand, and pressure will also fall. Any significant deviation from typical operating values is going to bring up alarms so you should see what is happening pretty quickly. If you look at conditions in the reactor building you'll be able to see a rising trend in radioactivity and humidity, so it won't take you long to work out that you've a Small LOCA on your hands—you know it isn't a large one because the primary circuit still has most of its pressure.

19.2   The Operator's Choice

You can't repair a Small LOCA with the plant running. Your procedures will probably get you to isolate CVCS charging and letdown flow, just in case it's one of these lines that is leaking, but to be honest, the chances are that your plant is heading down in pressure and you won't be able to stop it. So you have a choice. Do you watch things happen, doing nothing until reactor trip and safety injection setpoints are reached (on, for example, low primary circuit pressure)? Or, having diagnosed the fault, do you step in to manually trip the reactor and initiate a Safety Injection (SI)?

The safety case will assume the former—people who write safety cases are reluctant to claim any kind of operator intervention within 30 min of a fault starting. You can probably see that this is conservative, as it forces the designers to build a plant that can cope without the operators doing anything to help. More realistically though, your training in the simulator, together with your procedures, will put you in a position where you know that the best thing to do is to trip the plant (using the big red button) and then manually initiate an SI. Of course, it's just possible that you've read the indications and alarms incorrectly and so have misdiagnosed the fault… So what? A trip and an SI won't do any long term harm.

I mention this as being a choice because different countries train their operators in different ways. In France, it's typical for the operators to be trained to wait for automatic setpoints to be reached. In the UK, it's more common for the operators to be asked to take pre-emptive actions. There are advantages and disadvantages to each approach, but there isn't a significant difference in risk. Ultimately, it's the automatic systems that provide the backstop, regardless of operator actions. If you visit a nuclear plant in a different country, don't assume that the operators will have been trained in precisely the same way as you!

19.3   Finding a Balance

In a Large LOCA, safety injection is all about getting water back into the core and keeping it covered, but this is a Small LOCA. The core hasn't uncovered, so what good is an SI signal?

Think back to the centrifugal pump pressure/flow curve that you saw in the last chapter. With all of your ECCS pumps starting you'll have one of these curves for the high pressure pumps and another one for the low pressure pumps. The chances are that the pressure won't have fallen far enough for the low pressure pumps (or the safety injection accumulators) to be able to provide any water, so let's concentrate on the behaviour of the high pressure pumps.

At the start of the Small LOCA, the primary circuit pressure will be too high for the high pressure pumps to deliver any water. But the primary circuit pressure will fall as the pressuriser water level falls. The pressuriser heaters will try to combat this, but their effect is limited, and they will turn themselves off when the water level in the pressuriser falls too far (they'd be damaged if they were running in steam). Also, don't forget that when you trip, the drop in Thot will cause a dramatic drop in pressuriser level, reducing pressure still further. The upshot is that, with a Small LOCA, the primary circuit pressure will eventually fall below the pressure at which the high pressure ECCS pumps can start to inject water.

As the primary circuit pressure falls further, this injection of water will increase, and the leakage flow will decrease as there will be less pressure pushing it through the breach in the primary circuit. Eventually, you'll reach a point where the injection flow matches the flow of water out of the leak and pressure will stop falling (Fig. 19.1). Note that the ECCS Injection flow curve is wider than the one you saw in the previous chapter because it represents the flow from four high pressure pumps, rather than just a single pump. As an operator, you haven't had to find this flow balance, the physics and engineering do it for you, but this is the first step in dealing with a Small LOCA.

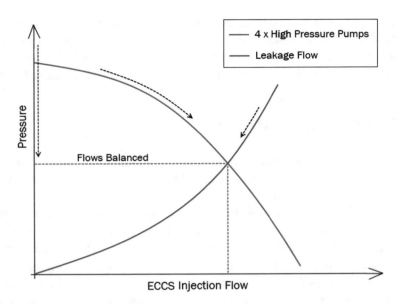

Fig. 19.1 ECCS injection and leakage flow balance

19.4   Moving On, Moving Down

Your plant is now stabilised, but it's hardly a sustainable position. You have a leak from the primary circuit, and although you're maintaining inventory for the moment, you might not be cooling the core very effectively. Some heat will be being removed by the water leaking from the primary circuit, but if the core heats-up it could cause boiling in the primary circuit and become uncovered.

What you need to do is move your primary circuit to a depressurised, cooled state. The automatic systems won't do this for you, so here's where your training is going to be important.

Firstly, look at the pressuriser water level—do you even have one, or was too much water lost before the ECCS pumps reached that balance point with leakage flow? What you'd really like to do before you do anything else is to get more water into the primary circuit, but how? Look at Fig. 19.1 again. Imagine if you could lower the pressure in the primary circuit below the balance point, what would happen? Well, ECCS injection flow would increase, and leakage flow would reduce. You'd have a net inflow of water into the primary circuit, and pressuriser level would start to rise.

Perhaps surprisingly, you still have a degree of control over primary circuit pressure: you have pressuriser sprays. If you open the spray valves, colder water will be sprayed into the pressuriser steam bubble, condensing some of the steam, so pressure will fall. Don't be confused into thinking that the sprays are 'adding' water—they aren't, as it's being taken from the cold legs—but you will see an increase in ECCS injection flow and pressuriser water level as primary circuit pressure is reduced. When you've got enough water in the pressuriser, you can turn off the sprays and return to your balance point. Incidentally, make sure the pressuriser heaters don't turn themselves back on as the water level rises; you don't want the pressure to go up!

OK, so now you've got a reasonable level in your pressuriser. Next, have a look at temperatures and pressures in your primary circuit. You should be well below boiling point, even at this reduced pressure—a 'sub-cooled margin'. If not, try dumping a bit more steam from your steam generators to bring temperatures down.

Why are the pressuriser level and sub-cooled margin important? Because you're now going to be doing something brave… you're going to start shutting down ECCS pumps. Figure 19.2 shows you what happens when you shut down the first high pressure pump. Your injection flow will reduce (by roughly 25%), so pressuriser level and pressure will initially fall as you'll have a net loss of water from the primary circuit. While it's lower, your injection flow is still significant. Be patient. It might take a few minutes, but you will find that the plant moves to a new balance point, with a lower primary circuit

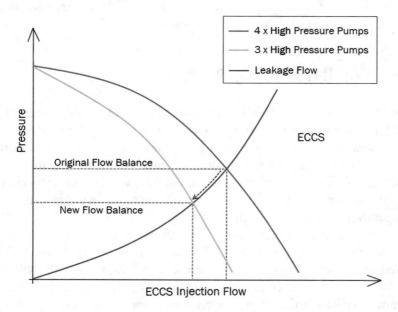

Fig. 19.2 Shutting down a high pressure ECCS pump

pressure and a correspondingly lower leakage flow (matching the new ECCS flow). You've made progress.

Now do it all again. Use sprays to recover pressuriser level and steam dump to restore sub-cooled margin, then shut down a second high pressure pump. Then repeat for the third pump, so that just one high pressure pump is in service. When you shut down the last pump, there'll be a more significant drop in pressure, all the way down to the pump curves for the low pressure ECCS pumps. You'll need to allow for this with a higher pressuriser level and sub-cooled margin, but your procedures will specify the values you need.

Incidentally, if you've started to worry about the cooldown giving you positive reactivity, there's no need. The water that the ECCS pumps are injecting is heavily borated (2500 ppm from the refuelling water storage tank) so the core will stay nicely sub-critical.

Go through the same process for the low pressure pumps and eventually you'll have all of the ECCS pumps shut down, and the leak flow will be matched simply by the CVCS charging. Surprising? It shouldn't be: a leak rate of a few litres per minute at 155 bar primary circuit pressure will reduce to a few drops per minute when the primary circuit is depressurised.

Well done. You've just taken your plant through one of the most complex fault recovery procedures that operators have to be able to perform. Small LOCAs are not everyday events, but they have happened, especially at the Very Small end of the size-spectrum, so it's not surprising that you spend so much of your time training to deal with one.

19.5   Small LOCA, Big Problem

Three Mile Island Unit 2 (TMI2) was commissioned in 1978. It was a PWR, albeit an earlier design than yours. Some of the safety systems were simpler, and the design of Steam Generators (SGs) was markedly different from the ones you're used to, having vertical 'once-through' tubes rather than 'U-tubes'. This design of SG allows for better steam conditions in the secondary circuit—even some superheat—but the SGs run with less secondary side water in normal operation, so can dry-out more quickly.

Just a year after commissioning, TMI2 suffered a trip from power due to a feedwater problem. Unfortunately, the auxiliary feedwater system had been isolated for maintenance—a breach of the operating rules under which the plant was running. The complete loss of feedwater to the SGs led to a rapid dry-out, after which the primary circuit was not being cooled. Decay heat caused the temperature and pressure in the primary circuit to rise rapidly, at which point one of the pressuriser's safety relief valves (a pilot-operated relief valve) lifted to relieve the pressure.

Here's where things started to go very wrong. The pilot-operated relief valve stuck open (a mechanical fault), but the operators had an indication in the control room suggesting that it was closed; in fact this indicator merely indicated that the valve had been sent a signal to close (you might remember that this was mentioned in the earlier chapter that discussed control room design). The operators now had a Small LOCA—from the top of the pressuriser—and didn't know it.

Pressure in the primary circuit fell. ECCS pumps started, but the loss of pressure was significant enough to lead to boiling in the core. One of the strange things about a LOCA at the top of the pressuriser is that it can cause pressuriser level to rise. Initially, such a LOCA is only leaking steam, not water, so there's no dramatic reduction in inventory. But as the pressure falls, boiling can begin in the core. This is what happened at TMI2. The steam bubbles in the core displaced water which flowed up into the pressuriser, causing its water level to rise.

This is when the operators (or the people that trained the operators?) made their biggest mistake. They didn't check the primary circuit sub-cooled margin so didn't realise that boiling was happening in the core. They assumed that the rising water level in the pressuriser was due to too much water being injected by the ECCS pumps, and they turned them off. This was despite other indications from within the reactor building pointing towards an ongoing LOCA.

With no injection of cold water, things deteriorated rapidly. Steam bubbles in the primary coolant water caused cavitation (vibrations due to bubbles) of the reactor coolant pumps, so these too were shut down. The flow provided by the RCPs, although full of steam bubbles, had been keeping the core covered. Shutting down the RCPs meant that the core was essentially uncooled and began to melt. More than a third of the fuel melted and ended up in the lower part of the reactor pressure vessel.

The error with the pilot-operated valve was eventually realised, and an intermediate block valve was closed, stopping the LOCA. Water was injected into the primary circuit using low pressure pumps, and the RCPs were eventually turned back-on, restoring core cooling. Thankfully, both the RPV and the reactor building remained intact despite a later ignition of hydrogen within the building. There was a minor leakage of radioactivity to the atmosphere sometime after the event, but there was no significant radiological release.

TMI2 was a wake-up call for the nuclear industry, especially in the USA. Coincidentally, the event occurred only 2 weeks after the release of the film "The China Syndrome", purporting to show a narrow escape from a 'meltdown' event at a nuclear power station. TMI2 showed how maintenance errors, inadequate operator training and poor instrumentation design could come together to write-off a nearly-new reactor. Operator training quickly improved, and more modern plants incorporate design changes that make such an event far less likely. At some plants, for example, the operators are unable to shut down ECCS pumps within a set period following an SI signal. From a strictly nuclear safety perspective, TMI2 could be seen as a vindication of PWR design—a melted core, but with almost no release of radioactivity to the environment.

19.6   Fault 4: Steam Generator Tube Leak (SGTL)

A Steam Generator Tube Leak (SGTL), or Rupture (SGTR)—the rupture is simply a larger leak—is not really a distinct fault. It's a kind of Small LOCA. But there's a good reason for including it in this chapter. An SGTL is a potential 'Containment Bypass fault'. The primary circuit is at a much higher pressure (155 bar) than the secondary circuit (70 bar), so any tube leakage will be from the primary circuit to the secondary.

Primary circuit water will be radioactive. It will contain small amounts of tritium (hydrogen-3) together with a range of dissolved gases and corrosion products, which may have become radioactive in passing through the core. In normal operation, this isn't a problem as the water is contained within the primary circuit, but an SGTL can release this radioactive water into the steam and feedwater systems.

Radioactive dose rates might then rise above background levels in normally 'clean' areas such as your turbine hall. More significantly, especially if it's a significant leak (or rupture), the pressure in the affected SG will rise to a level where the main steam safety valves (MSSVs) or power operated relief valve (PORV) could open. You'd then find that you were discharging radioactive steam directly to the environment. This is what we mean by a 'Containment Bypass Fault'.

As an operator, how you deal with an SGTL will affect how big a radioactive release you have, and for how long. So again, you'll spend a lot of time training on these on the simulator. Everything you've seen in this chapter on tripping, initiating a safety injection and driving the plant down to a low pressure/temperature condition still applies; arguably with a bit more urgency.

But there's a difference—the leakage flow is into the SG (secondary side) rather than into the reactor building. That means that what you're aiming to do is to drop primary circuit pressure to match the pressure in the secondary side of the affected (leaking) SG. If you can do that, the leakage flow will stop.

There's actually a trick to doing this on a large PWR, though most people struggle to understand it at first sight. To begin with, you need to identify which SG has the leak. This might be apparent from a rising water level, or a drop in feed flow, but it's more likely that the fault's been revealed to you by multiple radiation alarms from monitors attached to the secondary circuit. Once you have this information, isolate the affected SG by closing its main steam isolating valve (MSIV) and its feedwater valves. Now the trick: dump lots of steam from the other SGs to pull the temperature and pressure of the primary circuit down quickly. You'll have to keep an eye on your core's shutdown margin (you may have to add boron to stay subcritical), but that should be manageable.

If the cooldown is fast enough, it will cool the water in the lower part of the affected SG (around the tubes). The water towards the top will stay hotter, keeping the steam pressure in that SG high; think of it working a bit like the pressuriser. We would describe this SG as being 'Stratified' meaning that different layers of water are at different temperatures. With no flow of water or steam in or out, there's nothing to disturb the stratified layers. By keeping the pressure high in the affected SG (at around 70 bar), you can move the primary circuit pressure down to match it, rather than everything coming down together.

If you can manage this fast cooldown, you'll stop water leaking across the tubes much more quickly than if you wait for the primary circuit to be fully depressurised. Figure 19.3 shows how the leaking SG compares with one of the others once you've achieved this 'temperature stratification'. Notice how the stratified SG has no feed flow and has stopped producing (new) steam. In contrast, although at a lower steam-space temperature, the SGs used for cooldown are still being supplied with feedwater and are producing steam.

Figure 19.4 shows the conditions that you are aiming for in the primary and secondary circuits.

When you've achieved stratification in the leaking SG, and your target conditions in the other SGs and primary circuit, you will have stopped any radioactive water leaking into the affected SG or being released to the environment. Now follow your procedures, cool down the plant and repair (or plug) the leaking tube. You can expect to be shut down for a few weeks.

(a) (b)

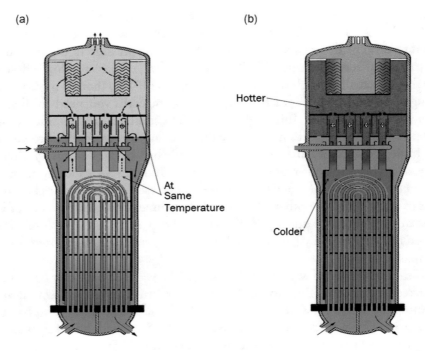

Fig. 19.3 Steam generator temperature stratification (**a**) steam generator used for cooldown, (**b**) stratified steam generator

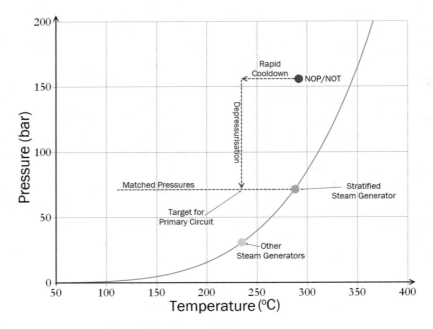

Fig. 19.4 Target conditions following an SGTL

19.7   How Is That Acceptable?

You might be wondering how it is that PWRs are allowed to operate if they can undergo an SGTL, i.e. a fault which releases even this small amount of radioactivity into the environment? Part of the answer is that SG Tube Leaks were relatively common, in the past. With more than 5000 tubes in each SG, there are a lot of potential leak sites. However, older PWRs used different SG tubing material from newer plants, and the industry has learnt a lot about SG chemistry and SG tube inspection techniques over the years, so a more modern plant would expect to run for many decades—possibly for the whole of its operating life—without an SG Tube Leak.

The other part of the answer is to think back to the chapter on Safety earlier in this book (Chap. 17). You'll remember that Licencing of your PWR depends on you having and following a set of 'Operating Rules'. Well, one of these rules will be an upper limit on how radioactive the water is in your primary circuit. If you're below this limit, then the modelling done to support your safety case will have shown that the release of radioactivity from an SGTL will be small enough to be tolerable. If you're above the limit—say, through multiple fuel failures while running at power—then you'll have no choice but to shut down. The limit on primary circuit radioactivity in your operating rules is a direct link back to the analysis of faults. Looked at another way, in complying with this operating rule you are showing that you are always ready for an SGTL to occur.

20

What Else Can Go Wrong?

I want to start this chapter by showing you how the things that keep your reactor stable can themselves become a problem during a fault.

20.1   Fault 5: Main Steam Line Break (MSLB)

You can think of a 'Main Steam Line Break (MSLB)' as being the most significant example of the secondary side equivalent of a LOCA. Smaller breaks and leaks of feedwater and steam piping can clearly occur, but as with the Large LOCA, I'm going to describe the big fault first.

The Main Steam Line (MSL) pipework on your plant is large. Each of the four pipes is around 0.8 m in diameter, carrying half a tonne of steam per second at 60 mph. So what happens if one breaks? Well, of course, there'd be a release of steam from the pipe, but it's what happens to the associated Steam Generator (SG) that turns this into an interesting fault.

As you've already seen, the SGs are saturation devices—that is the secondary side of each steam generator sits on the boiling curve with the temperature and pressure tightly coupled. This contributes strongly to the stability of the plant in day-to-day operation (see Chap. 12).

But if an MSL were to break, the pressure in the affected SG would be free to fall. Actually, that's not entirely true; there's a flow restrictor in the top of each SG that limits the rate of pressure drop somewhat, but the pressure will still fall rapidly. As the pressure decreases, the water in the SG will boil vigorously and, at the same time, cool down very quickly. The secondary side of the SG would slide down the boiling curve, cooling the SG tubing as it went. The

© Springer Nature Switzerland AG 2019
C. Tucker, *How to Drive a Nuclear Reactor*, Springer Praxis Books,
https://doi.org/10.1007/978-3-030-33876-3_20

tubes would cool the primary circuit water in that loop, meaning that Tcold in that loop would fall very quickly. Even allowing for mixing of the water between the loops before it enters the core—though this isn't really very good mixing—the reactor core is going to suddenly see much colder water. In PWR terms, this is a 'Cooldown Fault'.

Colder water acting through the (strongly negative) Moderator Temperature Coefficient (MTC) will inject positive reactivity. This will cause reactor power to rise rapidly. Incidentally, if the fault is sustained for long enough, the control rods will initially drive out of the reactor—in response to the falling temperature—thereby making the power rise even steeper. An uncontrolled cooldown fault on a PWR risks fuel damage by being a significant 'overpower' fault, and that's why an MSLB is of interest.

In reality, in the event of an MSLB, your PWR will automatically trip—probably before you've registered the first alarm. You'd expect it to trip on a range of parameters including reactor power (too high), Tcold (too low), steam line pressure (too low), SG water level (possibly too high to start with, then too low), reactor building pressure (too high) etc. Even so, for some of your operating cycle, the initial increase in reactivity might be high enough to overcome the negative reactivity inserted by the control rods. In other words the reactor could return (briefly) to criticality until sufficient boron had been injected to shut it down.

I've been talking about the other SGs as if they are unaffected by the MSLB. That's only true if they can be isolated quickly from the broken pipework. If the break is between an SG and the main steam isolating valves (MSIVs, see Chap. 10), then this is easily done. Closure of the MSIVs isolates the three unaffected lines from the broken one. If the break is downstream of the MSIVs, then their closure isolates all four SGs from the break, so terminates the cooldown entirely. Bearing in mind how quickly this needs to happen (seconds), it won't be a surprise to you to know that the Reactor Protection System (RPS) on your PWR will close these valves automatically if it detects a sudden drop in steam pressure or Tcold in any of the MSLs or SGs. These same indications will prompt an automatic reactor trip and a safety injection signal, so you will be well on the way to achieving a safe shutdown state.

As with LOCAs, there will be many other possible, but much smaller (and more likely), pipe leaks on the secondary system that could lead to a cooldown of an SG. A simple example would be the failure—to an open position—of one of the main steam safety valves (MSSVs, Chap. 10). In steam demand terms, this would represent an extra 5% in steam demand and so reactor power would rise by 5% to follow the steam demand. On your PWR that probably won't even cause an automatic reactor trip, as trip settings will be nearer 110% power.

With a stuck open MSSV, the operators would do the sensible thing: quickly reducing turbine load by 5%, bringing the total steam demand and reactor power back within limits. Then they'd work out how they were going to close the valve and/or shut down the plant. Even smaller examples of cooldowns might be caused by a drain on a steam line developing a leak, with the operators only having to move turbine power down by a few MWs to compensate. You should note here that brief overpowers are usually not a problem on a PWR. Most post-trip behaviour is governed by decay heat level, and so is dependent on power history much more than instantaneous pre-trip power level.

20.2   Fault 6: Severe Accidents

Now that you've become familiar with the textbook PWR faults, it's worth taking a moment to consider how bad things might become. This is the subject of 'Severe Accidents' and is characterised by core cooling being so poor that significant fuel damage is likely to occur (as it did at Three Mile Island 2).

There have been very few PWR Severe Accidents (only one on a commercial PWR), though the Boiling Water Reactors (BWRs) faults at Fukushima were somewhat similar. This means that much of the subject is based on experiments (at both small and large scales) and computer modelling. Having said that, there are lots of both of these, so there's a broad consensus on how a Severe Accident would progress.

By definition, a Severe Accident is one in which you've failed to adequately cool the core. How you've managed this on a modern PWR is anyone's guess. You'd need a large number of unrelated systems to develop a problem or be damaged at the same time. It's not impossible, but it'll be a very low likelihood compared with the faults you've already seen.

Firstly, the fuel: you've lost cooling, so the fuel is going to suffer. There are a whole range of physical and chemical processes at work when decay-heat-producing fuel is uncovered and uncooled. Some of these, such as the zircaloy fuel cladding reacting with steam, can actually make things worse by adding heat. Other effects, such as strong convection currents within the primary circuit, can remove some of the heat. Significant fuel damage doesn't happen straight away, so there are opportunities for operators to step-in and re-establish core cooling (if they can). Their procedures will highlight all of these. But we're talking here about severe accidents, so let's keep going…

If fuel cooling is not re-established, the core will begin to melt. Unfortunately, this makes it more difficult to re-establish effective core cooling as the geometry of individual fuel pins, and the gaps between them will have been lost. Melting will also liberate radioactive gases and volatile chemicals from inside the (failing) fuel pins. If this fault had started with a LOCA, then these radioactive fission products will be able to enter the reactor building. On the bright side (yes, there is one), this same loss of geometry means that criticality is very unlikely to be possible, even when you get water back into the vessel.

It is likely that some of the molten core will travel downwards to the bottom of the reactor pressure vessel (RPV). This material is usually called 'corium', as it's a mixture of molten materials from the core, usually at very high temperature. If there's a substantial amount of corium—at TMI2 there was around 20 tonnes—the heat could cause the RPV to fail. Experiments using simulated corium have shown that this can happen as a small hole that grows larger, or more dramatically, by a large tearing of the RPV.

So, let's take a step back. We started with three barriers to fission product release. The fuel/clad, the primary circuit and the reactor building. In this scenario, our unspecified 'Severe Accident' has failed the first and second barriers. So now, all of our efforts need to go into protecting the third barrier, the reactor building, as it's this that stands between just having a written-off reactor, and being responsible for a significant release of radioactive material. So what might threaten the reactor building?

Well, the original fault and the subsequent vessel failure may have given rise to high temperature and pressures within the building. But, as you've already seen, these buildings are large, strong and (to a great extent) empty. There's room for lots of steam to be released and to expand without the building being damaged. As an operator you can't control this—it's part of the design—but any cooling that you can bring to bear within the building will be beneficial. Eventually, you are going to have to find a way of cooling (or venting) the building to prevent it from failing through overheating or over-pressurising. There are many different ways of achieving this cooling worldwide.

The next hazard we might worry about is hydrogen. Hydrogen gas is produced by the reaction between zircaloy fuel cladding and steam. There are other routes for its production in a severe accident, some chemical, and some due to radiation. This hydrogen can be released into the reactor building where it will mix with air and could subsequently ignite. At TMI2, there was a burning of hydrogen some hours into the fault, causing an approximately 2 bar spike in reactor building pressure. Early PWRs had electrical recombiners to remove hydrogen in a fault; though it must be admitted that if you had power for these, you were less likely to be having a severe accident. More modern plants, including yours, have 'Catalytic Recombiners' that work passively without electrical supplies. This is why hydrogen is not considered a significant hazard on a modern plant.

Finally, there's a potential threat to the containment floor (the 'basemat') from the molten corium that's fallen out of the bottom of the RPV. Some modern PWRs have special cooled floor areas over which this material can safely spread. Others, more typically, rely on getting water underneath the vessel—'Containment Water Injection'. While this will increase the steam production within the building, experiments suggest that the violent boiling will break-up the corium into small, coolable, pieces and so prevent the containment basemat from being threatened.

20.3   Fukushima Daiichi

On the 11th March 2011, a very significant earthquake occurred off the coast of Japan; the most powerful yet recorded. The tsunami caused by the earthquake swept onto the coast of Japan's main island, tragically resulting in the deaths of nearly 16,000 people, with 2500 still missing.

There used to be six reactors on the Fukushima Daiichi site. These were all early Boiling Water Reactors (BWRs) dating from the late 1960s and early 1970s. Of particular note is the location of reactors 1–4, being quite close to the sea and to each other. Reactors 5 and 6 were separated from the first four reactors and set a little further back.

When the earthquake struck, reactors 1–3 were operating at power, while reactors 4–6 were already shut down (the fuel had been removed from reactor 4). Reactors 1–3 shut down automatically in response to the earthquake, as they were designed to do.

However, when the tsunami reached the site some 50 min later, it overwhelmed the sea defences leading to extensive flooding. This included the back-up generator buildings and other essential equipment for cooling and control of reactors 1–4. Cooling of reactors 1–3 failed, as did the cooling of the fuel storage pond above each reactor (including reactor 4). Sufficient equipment survived the flooding on reactors 5 and 6 to maintain cooling, so these won't be mentioned again.

The operators tried many different approaches to connecting generators and batteries to the cooling systems for the reactors, but the tsunami had damaged local roads, and this made transporting any additional equipment to the site very difficult.

The cores in each of reactors 1–3 underwent significant melting and damage, with the relocation of the corium into the lower levels of each reactor building. Unlike modern PWRs, the containment buildings of these early BWRs were not designed to cope with severe accidents. The pressure in these buildings rose, and it eventually became necessary to vent the containments to avoid their failure. Due to the extensive fuel damage, the vented gases were rich in hydrogen and caused explosions during the venting process. It was these hydrogen explosions, in the spaces above the reactor containments, which were seen worldwide in the following days.

The fuel pond above reactor 4 had not suffered as badly as had been first feared (it has now been emptied). Clean-up and decontamination of the Fukushima Daiichi site is progressing well but will take many years to complete due to the extremely high radiation levels in the vicinity of the relocated corium.

Much of the criticism directed at both the operators and regulators in Japan has centred on the 'reasonably foreseeable' nature of the tsunami and hence the inadequacy of the sea defences against natural hazards. Internationally the responses have included the closure of some older BWRs (e.g. in Germany) and more in-depth preparation for events considered to be just outside the design basis of other plants, especially those that might include failures of multiple plant systems.

20.4   In the Longer Term

As TMI2 and Fukushima have both demonstrated, it is possible to reach a safe, stable plant state, even if a severe accident has occurred. On a modern PWR, this can be achieved without a significant release of radioactivity to the environment. What then?

What comes next is 'clean-up'. Cooling is maintained while short-lived fission products are allowed to decay. The wastes that are left (which might include previously molten corium) are located, characterised, removed and packaged. It's expensive and time-consuming, but it is achievable.

20.5   The Best Way to Deal with a Severe Accident...

... is not to have one.

Build a well-designed plant with three or four safety trains and a modern, strong reactor building. Train the operators thoroughly, including on a wide variety of accident procedures. Be cognizant of all potential hazards, both internal and external, natural and man-made. Maintain the plant thoroughly and keep your eyes and ears open to international operating experience.

Always operate the plant as if a fault is just about to happen. That way, you'll stop any that do from developing into a severe accident.

21

When You Run Out of Oomph

As the fuel is burnt-up in day-to-day operation, your reactor will lose reactivity. To keep it running you'll slowly dilute some of the boron out of the primary circuit—perhaps 2 to 3 ppm per day, starting from around 1500 ppm (at full power). You might think that you could keep doing this all the way until you reach zero ppm, but as you've seen, in practice, you'll likely want to stop before that, so as not to have to use huge volumes of water to continue diluting (refer to Chap. 12). When you reach that point you are ready to refuel your reactor; you're ready to start your next 'Refuelling Outage'—an 'outage' is just another word for a period that you're shut down.

21.1 Coast-Down

There is another way of clawing back some reactivity—drop power and gain reactivity from the power defect. Typically, you'll find that you can offset the loss in reactivity from burnup if you reduce power by around 1% per day. We call this 'coast-down'. From a reactor operator's perspective it's really easy to do. You simply stop the (by now quite large) dilutions, and instead, you reduce power on your turbine by a few MW every few hours. With such small changes in power, xenon transients are unlikely to be a problem, and axial flux difference will be easy to predict as everything changes so slowly.

On the other hand, if you're going to coast-down, your station will be losing output, so this probably seems an odd thing to do? But, imagine that you've loaded your core at the start of the cycle with an assumed date for the next refuelling outage—say, 500 days away. What happens if you get most of

© Springer Nature Switzerland AG 2019
C. Tucker, *How to Drive a Nuclear Reactor*, Springer Praxis Books,
https://doi.org/10.1007/978-3-030-33876-3_21

the way through the operating cycle and then have to move your next outage date back by a couple of weeks, perhaps because of the unavailability of key contractors. How do you keep the plant running for another 14 days? That's when coast-down would be a sensible option. Even at the end of 2 weeks, you'd still be generating 85% of your full output; much better than shutting down for 2 weeks waiting for the contractors to arrive.

Some plants even design their cores with a planned period of coast-down each cycle as this can increase overall fuel burn-up; when you're paying so much for each fuel assembly, it's good to get as much as you can out of each one (provided you stay within your safety case limits on burn-up).

21.2   Shutting Down

Shutting down the reactor for an outage is just another power change. So you tackle it in the same way as any other power change. Set the turbine load to slowly decrease and borate the primary circuit to offset the reactivity you'll get back from the power defect. It'll feel a bit strange as you've probably spent months diluting to hold power steady and now you're going the other way!

Power changes can be a bit tricky as you get close to the end of an operating cycle. The Moderator Temperature Coefficient (MTC) will be very negative, so a small change in temperature (power) will need to be offset by a lot of boron. Your control rods may not give you as much negative reactivity per step as they did at the start of the cycle. The fuel assemblies in which the control rods are located may now be running at a lower power relative to other assemblies, so the rods can affect less of the neutron flux. Finally, xenon can be a bit of beast at the end of the cycle. Different concentrations of xenon can develop in the top and bottom of the core, leading to xenon transients that move around in your reactor! None of this will be a big problem if all you are doing is shutting down, but it may make the reactor more difficult to stabilise at lower powers. So once you start driving it down, keep going...

When your turbine is at a low enough power, you'll disconnect it from the grid and allow it to spin-down. Your reactor will still be running at low power, dumping steam, and you can either drive the control rods in to reduce power all the way or simply use the big red button to trip the rods in. Just like after an unplanned trip, your first aim is to stabilise the plant at Normal Operating Pressure and Temperature (NOP/NOT).

21.3   Cooling Down

Have a look at the boiling point curve in Fig. 21.1. This time I've marked the NOP/NOT temperature and pressure (there'll be only a small difference between Thot and Tcold with the reactor shutdown). That's where you are now. Where you need to get to is right down on the left, i.e. zero pressure and cold enough for you to get on with your outage.

How do you get there?

You can't simply drop the pressure, or you'd end up on the boiling curve—the water in the primary circuit would turn to steam! Similarly, you can't just cool-down, or you'd risk the integrity of the primary circuit—colder steel loses strength. Instead, you're going to need to bring temperature and pressure down together, something like the dashed line I've shown in the figure. It's not as hard as it sounds because you can control pressure and temperature independently.

Control of primary circuit pressure is through the pressuriser. If you turn off all of the pressuriser heaters, the pressure will slowly fall as the pressuriser cools down. That's probably not going to be fast enough for you, so instead think about using the pressuriser sprays. The more the spray valves are opened,

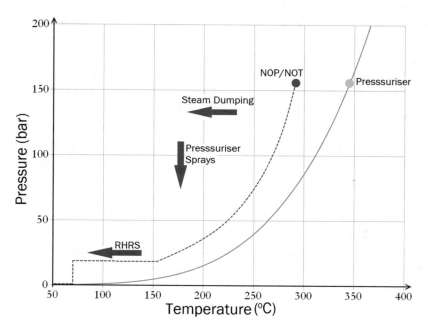

Fig. 21.1 Cool-Down Curve

the faster pressure will fall—remember that the water for the sprays comes from the cold legs, so, at least to begin with, it's a good deal colder than the steam bubble in the pressuriser.

Primary circuit temperature is controlled in just the same way as it is at power, using the steam pressure on the secondary side of the steam generators. Your turbine is shut down but you can still dump steam to the turbine condensers (or to atmosphere via the PORVs, if the condensers aren't available). Initially, the pressure at which this steam is being dumped will be around 75 bar. That will hold steam temperature, and hence Tcold, at just over 290 °C—the NOT part of NOP/NOT. To cool down the primary circuit you adjust the steam dumping pressure down a little bit. This will increase the amount of steam dump, reducing SG pressure. As the SGs are saturation devices, this will also reduce the secondary side temperature, which will, in turn, reduce the temperature of the water returning to the reactor (Tcold).

Now the clever bit: if you reduce steam dump pressure gradually and open the pressuriser spray valves, then you can bring primary circuit temperature and pressure down together. This will keep you well away from the boiling line and ensure that you don't overstress the primary circuit steel. Interestingly, although Tcold and Thot (close together) will follow something like the dashed line in Fig. 21.1, the pressuriser will follow the boiling curve down—it still has a steam bubble—you don't have to control this, the physics does it for you. As the pressure reduces, water in the pressuriser flashes to steam, cooling as it does so. The steam bubble stays in equilibrium with the pressuriser water.

I'll be honest. This is a little more difficult than it sounds. If you want to achieve a constant rate of cooldown, say 25 °C/h, this doesn't match a constant rate of fall of steam pressure on the secondary side—it's a boiling curve, not a boiling straight-line. Additionally, the effect of the pressuriser sprays varies with the temperature difference between the pressuriser and Tcold. On older plants, this was a difficult manoeuvre which kept the operators very busy!

You have a modern PWR. You can select the desired rate of cooldown on your steam dump system, and the automatic control systems will work out what that means in terms of steam pressure changes. Similarly, if the control systems incorporate something like the dashed line in Fig. 21.1, they can control pressuriser spray to give you the desired pressure, wherever you are in the cooldown. Set this going automatically, and you can reach a cooled, depressurised primary circuit in 8 hours, or so.

21.4   Reactor Coolant Pumps

What would you do with your RCPs? They add heat to the primary circuit, 5 MW or more each, so it doesn't make a lot of sense to keep them in service while you're trying to cool down. On the other hand, they circulate the coolant very well, improving the transfer of heat to the SGs and ensuring proper mixing as you change boron concentration. The compromise is probably to be to shutdown three of your four RCPs. This minimises the heat input while still giving you good circulation of coolant; forwards through the loop with the running RCP, backwards through the other three. It sounds odd, but it works fine. Choose the running RCP carefully: it needs to be in one of the loops which supply pressuriser sprays (remember that it's the pressure from a running RCP that drives the spray water from a cold leg up to the top of the pressuriser).

There's a little complication with running RCPs, and that's the seal. It turns out that the seal could be damaged if the RCP is running with very low primary circuit pressure. In normal operation the pressure in the primary circuit pushes upwards on the reactor coolant pump impellor, and the shaft and the RCP seals have been designed to work with this in mind. As the primary circuit pressure falls below around 20 bar, the upwards force on the RCP is insufficient, and its weight could push down on and damage the seal. This is why the dashed line in Fig. 21.1 comes down to 20 bar then holds steady until the temperature is low enough to shut down the last RCP. Once it is shutdown, the final bit of depressurising and cooldown can take place.

21.5   Boron

You remember that you were adding boron as you were reducing power. This was to offset the power defect, i.e. the increase in reactivity as the reactor power was reduced. Well this needs to carry on even when the reactor is shutdown. The MTC and FTC stay negative most of the way down, so reactivity will increase as you cool down the primary circuit. At the end of an operating cycle this is unlikely to be a problem, but you're going to be refuelling the reactor in a few days, massively increasing its reactivity. If you don't do the refuelling at a high enough boron concentration, your cold reactor will go critical with all the control rods in and the lid off…

So, even though the reactor is shut down, keep borating. A modern plant is likely to need around 2500 ppm of boron to ensure that the reactor stays subcritical by an adequate margin (say, 5 Niles) during and after refuelling. This isn't so difficult: you'll need to add lots of water to the primary circuit as it cools down (more than 50 tonnes) to compensate for the shrinkage of the water inside it as temperatures fall. Just add the boron to this make-up flow.

21.6   The Chemists Are in Charge

In normal operation, the primary circuit is heavily dosed with hydrogen to keep out free oxygen, and so minimise corrosion. Now you have a problem: if you break into the primary circuit when it's cold, that hydrogen will come out of solution and could mix with air to cause a fire or explosion. You need to get rid of the hydrogen and replace it with oxygen—you don't want a hydrogen explosion. At low temperatures corrosion is less of a concern, so this is a reasonable change to make. But how?

By using the volume control tank (VCT) in the CVCS. Hydrogen is usually introduced into the primary circuit water at the VCT, so now the chemists remove the hydrogen from this tank and replace it with an inert gas (nitrogen). As the hydrogen level in the primary circuit falls, the chemists can speed-up the changeover to oxygenated conditions by adding chemicals such as hydrogen peroxide that break down to release oxygen. This is another good reason for keeping that last RCP running for a day or so—ensuring mixing of the primary circuit water until the chemistry changeover is complete. Eventually the chemists will be able to tell you that it's safe to shut down the last RCP and open up the primary circuit. You can't move forward until they do.

On older plants, this changeover from hydrogen to oxygen in the primary circuit water led to a lot of radioactive corrosion products dissolving in the water: the primary circuit water would suddenly become very radioactive, something the Americans called a 'crud-burst'. On more modern plants like yours, with good chemistry control and modern materials, the crud-burst is barely measurable.

21.7   Cooling When Cooled Down

You now have a relatively cold, depressurised primary circuit and secondary circuit. But if nothing is hot and boiling, how are you removing the decay heat? The answer is the Residual Heat Removal System (RHRS) shown in Fig. 21.2.

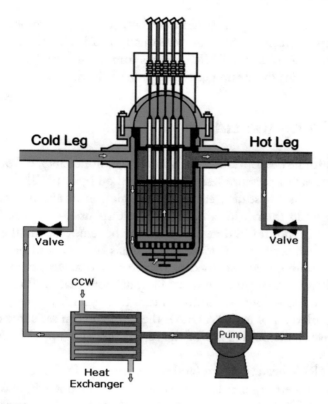

Fig. 21.2 Residual heat removal system

It's a pumped system, taking water from the four hot Legs and returning it to the four cold Legs, from where it flows through the core, back to the hot Legs—it doesn't flow through the steam generator tubes. On its way through the RHRS, the primary circuit water is cooled by a closed-loop cooling system, called the Component Cooling Water System (CCWS).

The CCWS cools all of the equipment important to nuclear safety on your power station including the reactor building, your safety injection pumps, control room, protection system computers etc. At modern plants, it's commonly divided into two or four separate trains, just like other safety systems and will be backed-up by essential diesel supplies in the event of a loss of grid. The CCWS will be cooled in turn by some kind of open-loop cooling system such as a sea-water system or an air-cooled radiator (or both, if you're lucky).

The RHRS is not designed to remove heat from your core at full temperature and pressure, but you probably opened the valves separating it from the primary circuit and started using it about half-way through your cooldown procedure. For a while, you would have been using it in combination with steam dumping from the SGs, but eventually you'd have stopped the steam dumping and just used the RHRS. This is important for what happens next as from here on in, the steam generators won't help you.

21.8 Lifting the Lid

Think back to the description of the primary circuit in Chap. 6. You'll remember that the reactor pressure vessel head is held on by bolts? There are around 50 of these bolts (or studs), each one weighing in at about a quarter of a tonne. You undo them with a huge hydraulic spanner—or Stud Tensioner/De-Tensioner. Once this is done, the bolts can be removed, and the RPV head lifted. The head weighs 150 tonnes, but that's ok; you still have the crane inside the building that was used to construct the power station—it previously lifted the reactor pressure vessel itself at more than 400 tonnes, so you know it can comfortably lift the RPV head. In Fig. 21.3 you can see an RPV head being lifted off of its RPV. Under the head you can see some of the control rod drive shafts poking out of the top of the RPV upper internals (refer to Chap. 6).

With the RPV head removed (and put to one side in the reactor building), you now do something that looks very strange to anyone who hasn't worked on a PWR (or BWR). You flood the stainless steel cavity above the reactor (the

Fig. 21.3 RPV head lifting

refuelling cavity, empty of water in Fig. 21.3) with more than a thousand tonnes of borated water. This water comes from the refuelling water storage tank (RWST), as this water isn't needed for safety injection now that the primary circuit is cold and depressurised. Everything you do with the reactor from now on happens underwater.

Water's great. It takes away heat from the fuel. It shields you (the operator) from the radiation coming from the fuel, and (best of all) you can see what you're doing through the water! If this last one sounds trivial, try working on a different design of reactor where fuel handling is done remotely behind thick shielding and with minimal instrumentation…

Figure 21.4 shows a flooded PWR refuelling cavity. In this picture, the upper internals have been removed from the reactor pressure vessel and are stored underwater in the cavity. You can see the tops of the fuel assemblies in the core, some of which have already been removed. You can also see the openings for two of the hot legs. Note that the control rods have already been unlatched from their drive shafts and so were left behind in the fuel when the upper internals were lifted.

Fig. 21.4 A flooded refuelling cavity

21.9   Defuelling, Shuffling, Refuelling

Most PWRs remove the whole core to the fuel storage pond next to the reactor building, every outage. This might seem an odd thing to do as we'll only be changing around a third of the fuel for fresh assemblies, but there are two good reasons for it. Firstly, removing all the fuel from the reactor building allows you to shut down and maintain all of the safety systems associated with the reactor—you just need to keep the ones in service that you need for the fuel storage pond. Secondly, and probably more importantly, the fuel that is staying in the reactor will be moving around to different locations. It'd be quite a juggling act to do this inside a partially loaded reactor vessel; especially as you'd also have to pick-up and move the control rods between assemblies so that they stayed in the correct locations. Some plants do it this way, but it's far more common to remove the whole core.

How do we move fuel assemblies? With a small crane (called a 'Refuelling Machine') that lifts one assembly at a time from the core and deposits it in a basket that then carries it (on its side) into the fuel storage pond. Once in the pond, it's placed in a vertical storage rack by another small crane. Typically, four or five fuel assemblies can be removed from the core in an hour, so a full core offload takes less than 2 days. Figure 21.5 shows one irradiated fuel

Fig. 21.5 Lifting an irradiated fuel assembly out of the core

assembly being lifted from a core. You notice the blue glow? That's Cerenkov radiation—an effect caused by placing something very radioactive in water (look it up, it's fascinating). To a physicist, that's an indication that the irradiated assembly is lethally radioactive, but all of this takes place underwater (at least 3 m of water) to provide cooling and shielding. You can stand on the side of the flooded refuelling cavity all day and get no recorded dose of radioactivity from the fuel.

Once in the fuel storage pond, the core components such as control rods can quickly be shuffled between fuel assemblies in the racks, without handling the assemblies themselves. The racks contain boron (as does the fuel storage pond water), so there's no danger of criticality even when control rods are removed. A full core shuffle only takes a day or so, as not every core component needs to be moved.

Refuelling is simply the reverse of defuelling, using the same cranes and basket, but remember that some of the fuel going back into the core is now fresh fuel. This will have been delivered in the months leading up to the outage and will already be waiting for you in the fuel storage pond racks. Some of your control rods might well have been shuffled into this fresh fuel after the offload. Refuelling tends to take a few hours longer than defuelling. Putting fuel into the vessel is more laborious than taking it out as irradiated assemblies tend to be a bit bent or bowed so have to be positioned carefully. The new core is much more reactive (around 20 Niles more!) than the old one so you'll be carefully monitoring neutron flux as the core is loaded, to ensure that it gets nowhere near criticality.

21.10 The Way Back

With the core reloaded, the upper internals are lifted in and the control rods re-latched to their drive shafts. The water, which will have been kept clean by a cavity clean-up system, is drained from the refuelling cavity (back to the RWST) and the reactor pressure vessel head replaced and bolted down onto the RPV. It sounds simple, but there's a good few days' of work in draining and cleaning the cavity so that it's spotless for the next outage.

Now you're going to get busy. All of those plant systems (including safety systems) that have been shut down and maintained during the outage have to be realigned and tested before being put back into service. A lot of that work will involve the operators in the control room. The outage is the only opportunity to do some of this work, so it's not surprising that even a lightly-loaded outage plan might contain more than 10,000 separate tasks.

The primary circuit is refilled with water (back-up into the pressuriser), pressurised to start the first reactor coolant pump, then the chemistry is switched back to hydrogen. Once this is done the heat-up can commence. Reduce the RHRS cooling and start all four RCPs. Again, with your modern plant you'll have automatic systems that keep pressure and temperature in-line with one another as you heat-up, but there's probably more testing to do on the way up so it won't be quick. As the pressure rises, you'll isolate the RHRS and switch over to steam-dumping. In a day or two you'll get the plant to NOP/NOT and can start to think about taking it critical. If the outage has gone to plan, all of the work on your turbine will be completed, and it'll be waiting for some steam.

21.11   Physics Testing

You might think that all you've got to do now is to follow your normal start-up procedures, as you saw in earlier chapters. You'd be wrong. Your new reactor is a very different beastie from the one you shutdown just a few weeks ago. It is much more reactive, and its MTC is much less negative (due to the high boron). Even the control rods will have a different effect on reactivity as they are in fresh fuel assemblies. What you need to do now is to treat this reactor as if you were commissioning a new plant. We call it 'Physics Testing'.

Physics testing starts by taking the reactor critical, but rather than doing this by withdrawing the control rods we're going to ask you to do it by slowly diluting the boron. Surprisingly, this is done with all of the control rods fully withdrawn, but that's OK as the boron is very high to begin with (its refuelling value, 2500 ppm in your case). Actually, the rods will already have been withdrawn once to test that they all drop in quickly enough when the reactor trip button is pressed—another test that can only be done in an outage with lots of boron.

Diluting to criticality is a slower process than rod withdrawal and is, arguably, a more cautious way of approaching criticality. You'll have a predicted critical boron concentration, but this will be based on computer modelling rather than on simple changes from the last criticality.

Once criticality has been achieved, the core characteristics such as MTC and control rod reactivity worths can be measured at very low power. As power is raised, this is followed by measurements of the power shape within the reactor. In just a few days, your physicists will be able to confirm that your core is behaving as expected and you can revert to your normal procedures.

21.12   Afterwards

It doesn't take long to get back to full steady power. If you've had lots of contractors on-site, they'll all have gone home, and the place will seem very quiet. You'll need to keep your eyes open for anything that might not be running quite right after being maintained, as this could force you to shut down again. A generally accepted measure for a successful outage is the plant running steadily for the next 100 days. Don't you believe it! 500 days is a good, post-outage run—all the way to the next outage.

You better start planning for it now.

22

Other Reactor Designs Are Available

This book is about Pressurised Water Reactors (PWRs). They are used to power two-thirds of the world's 450 nuclear power stations—with more under construction. But PWRs aren't the only reactor design out there. So, in this chapter, I'm going to (briefly) show you a few of your other options…

22.1   A Little Bit of History

At the beginning of this book, I introduced you to Chicago Pile 1 (CP-1). Clearly, CP-1 wasn't a power station, but it did demonstrate the feasibility of a man-made fission chain reaction using uranium as a fuel. I'm sure the scientists and military people involved in the project realised that a nuclear reactor could, at some time in their future, provide a compact, long-lasting source of energy, but that wasn't the project they were involved in. They were trying to make a nuclear bomb.

The scientists working on the American nuclear weapons program (the 'Manhattan Project') had realised that they could make a weapon in two ways. Firstly, they could take natural uranium out of the ground and enrich the U-235 content from 0.7% up to more than 80%. Secondly, they could use a natural uranium fuelled reactor to produce plutonium-239, which could be separated chemically from uranium, without using enrichment technology.

They pursued both options successfully. The weapon dropped on Hiroshima was a U-235 bomb ('Little Boy'), the one dropped on Nagasaki was a Pu-239 bomb ('Fat Man'). I'm not going to talk about nuclear weapons or their

© Springer Nature Switzerland AG 2019
C. Tucker, *How to Drive a Nuclear Reactor*, Springer Praxis Books,
https://doi.org/10.1007/978-3-030-33876-3_22

development anymore in this book—there are a lot of books on this subject already. What I want to talk about is what happened next…

I've called the Manhattan Project the 'American' project, but scientists from the UK and Canada were deeply involved, together with some who'd escaped occupied Europe. The UK and Canada probably expected the co-operation from the Manhattan project to continue beyond the end of the war, but they were to be disappointed. The Americans introduced legislation in 1946 (the McMahon act) that effectively shut their allies out of the nuclear program and prohibited sharing any further nuclear technology. This is the explanation for the different paths taken by the USA, Canada and the UK in developing their own pioneering nuclear technologies.

America had built enrichment plants, so was able to develop reactors using enriched uranium fuel. Enriching uranium allows a reactor designer to offset the loss of neutrons that occurs due to neutron capture by hydrogen atoms in normal (light water). America went on to develop mainly enriched uranium/light water reactors, such as PWRs and the similar BWRs (see below).

On the other hand, Canada had no enrichment plant, so had to build reactors fuelled with natural (unenriched uranium). This can't be done using light water but is possible using heavy water. Heavy water is water where most of the hydrogen is replaced with deuterium. Deuterium is a form of hydrogen that has an extra neutron, so is less likely to capture further neutrons. As it happened, one of Canada's significant contributions to the Manhattan Project was the construction and operation of a plant to produce heavy water using an electrolysis method. With these factors in mind, it made sense for Canada to concentrate on natural uranium fuelled/heavy water-moderated reactors—Canadian Deuterium/Uranium, or CANDU reactors.

The UK ended the war with no uranium enrichment plant and no heavy water production facility. From this, it was clear that early UK reactors would need to use a moderator with low neutron capture and that air or gas cooling would have to be used rather than water. This led to natural uranium/graphite-moderated reactors. Early UK reactors (e.g. the Windscale Piles, Chap. 17) were air-cooled and produced no electricity. Their sole purpose was the production of plutonium for weapons. The follow-on electricity-generating reactors used carbon-dioxide gas as a coolant; CO_2 being inexpensive, relatively inert and capturing far fewer neutrons than, say, nitrogen. It wasn't until the late 1980s that the UK took the decision to move to the, by then, much more common light water reactors (PWRs) and constructed Sizewell B.

Each of these reactor designs is described very briefly below. This should give you enough to get started if you want to research any of the other designs in a bit more detail. The numbers of reactors listed below are taken from the

International Atomic Energy Agency's 'Power Reactor Information System' (mid-2019 data) and exclude reactors that have now been permanently shutdown.

22.2 Pressurised Water Reactors (PWRs)

I don't need to describe PWRs in any detail as they've been the main subject of this book. The first PWRs were designed for use as power plants within American submarines. A nuclear-powered submarine has several advantages over a conventional (diesel-electric) submarine. These include the ability to stay submerged and hidden for days or weeks at a time and to travel long distances without refuelling. The use of PWRs in submarines allowed these to take-over from high altitude bombers in the American nuclear weapons program. Small PWRs have subsequently been used as power sources within a large number of submarines, aircraft carriers and ice-breakers.

In the late 1950s, a decision was taken to move one of these reactors onshore to generate electricity. This was the Shippingport Atomic Power Station, able to generate 60 MW of electricity from late 1957 onwards. Compare this with a typical modern PWR, able to generate 1200–1700 MW.

Following on from the Shippingport power station, a large number of PWRs were constructed in the US, and later in other countries including France, China, Japan and in Russia and Eastern Europe. Today there are around 300 commercial PWRs in operation, with another 40 under construction. That's excluding the hundreds of submarines and ships that have been PWR-powered since the 1950s. The PWR is by far the most common reactor type currently in the world.

22.3 Boiling Water Reactors

You can think of a Boiling Water Reactor (BWR) as a sort of truncated PWR. Rather than having separate primary and secondary circuits, the water is allowed to boil in the core producing steam that is then used to directly drive a steam turbine. Figure 22.1 shows the layout of a typical BWR.

You can see that a BWR has no Steam Generator tubes; instead, the steam drying equipment is located directly above the core. This means that the Control Rods can't enter the reactor from above as they would in a PWR, and instead have to be inserted from underneath. Rather than relying on gravity to give a fast shutdown, BWR control rods are pushed into the core by

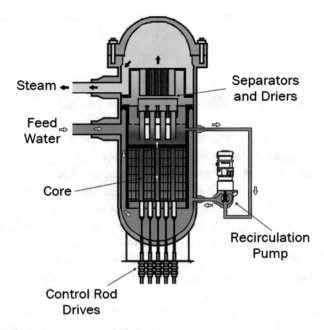

Steam

Feed
Water

Core

Separators
and Driers

Recirculation
Pump

Control Rod
Drives

Fig. 22.1   Boiling Water Reactor (BWR)

fast-acting pneumatic or hydraulic systems. BWR fuel is similar to PWR fuel; uranium oxide in a zircaloy cladding.

BWRs are surprisingly simple to control. The void coefficient (the reduction in reactivity with boiling, Chap. 9) is used to directly control reactor power. If you want more power, you circulate more water through the reactor pressure vessel, suppressing boiling deep within the core. If you want less power, you reduce circulation and allow boiling to pull reactivity and hence power down. To aid circulation of water, many BWRs have external recirculation pumps, as well as the supply of feedwater.

BWRs have tended to have lower costs of construction than PWRs as they have no steam generators, nor do they operate separate primary and secondary circuit pumps. However, you might remember that one of the hazards of the primary circuit was the production of nitrogen-16 as water flowed through the core. In a PWR this was primary circuit water so stayed in the reactor building. On a BWR this water—as steam—has to leave the building to drive the turbine. BWR turbines have to be heavily shielded to keep in the gamma radiation, and on-line turbine maintenance is severely restricted.

70 BWRs are currently in operation (mostly in the US and Japan) with just a few more under construction. The Fukushima Daiichi reactors 1 to 4 were

examples of BWRs, though arguably with fewer safety features than more modern plants.

22.4   CANDU Reactors

Heavy water is an effective moderator of neutrons while capturing far fewer than light water. A heavy water-cooled and moderated reactor, using natural uranium (non-enriched) fuel is therefore possible, and after a few years of experimentation, this approach was taken by the Canadians for their first nuclear power reactor in 1962.

CANDUs are 'Pressure Tube' reactors, with fuel rod bundles sitting horizontally in the tubes. The tubes sit in a tank of heavy water (known as the 'Calandria'), with coolant water being pumped through the tubes and out to Steam Generators that aren't dissimilar to those of a PWR. In fact, both the primary and secondary circuit temperature and pressure in a CANDU are pretty close to that in your PWR, and CANDUs are sometimes referred to by the name 'Pressurised Heavy Water Reactors (PHWRs)'. Up until the most recent designs of CANDU, the primary coolant water was heavy water as well. The latest models incorporate a small amount of uranium enrichment or use fuel composed of a mixture of natural uranium and plutonium, to enable light water to be used as the primary coolant. This restricts the use of expensive heavy water to only acting as the moderator in the calandria.

Interestingly, the pressure tube design allows for on-load refuelling of the reactor, by coupling fuelling machinery to either side of the reactor and pushing fuel along the horizontal channels. The water flow path in Fig. 22.2 appears complicated as water flows in opposite directions in adjacent fuel channels to even out the distribution of irradiated and fresh fuel.

Worldwide, CANDU reactors have proven to be a very successful design. Currently, 31 CANDUs are operational, a dozen of these being outside of Canada. In addition, India has 13 PHWRs based originally on the CANDU design. The UK built a similar experimental reactor—the Steam Generating Heavy Water Reactor (SGHWR)—but this ceased operation in 1990.

22.5   MAGNOX Reactors

The UK's first electricity-producing reactors were natural uranium fuelled, graphite-moderated reactors, using pressurised carbon-dioxide gas (CO_2) as a coolant. The uranium was used as a metal, clad in a magnesium can. The

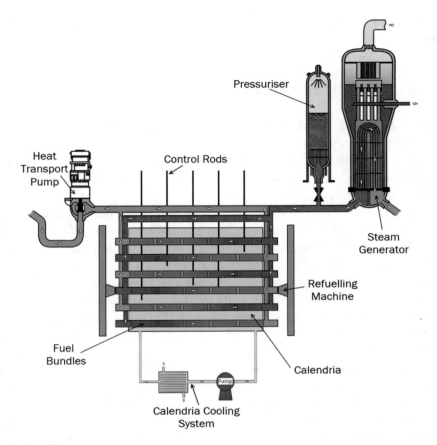

Fig. 22.2 CANDU Reactor

magnesium, which on its own, captures very few neutrons, had additives to reduce its likelihood of undergoing chemical reactions. For this reason, it was known as Magnesium Non-Oxidising cladding, hence MAGNOX.

A total of 26 MAGNOX reactors were built in the UK, including the first UK nuclear electricity generating station at Calder Hall (1956). The first MAGNOX plants (Calder Hall and Chapelcross) produced only small amounts of electricity as their primary purpose was to produce plutonium for the UK's nuclear weapons programme. All of the other UK MAGNOX are considered to be 'Civil' reactors and were not used to produce weapons-grade plutonium.

A diagram of a typical (steel-pressure vesseled) MAGNOX is shown in Fig. 22.3. The last four MAGNOX reactors (at Wylfa and Oldbury) had concrete pressure vessels, not dis-similar to Advanced Gas-cooled Reactors (see

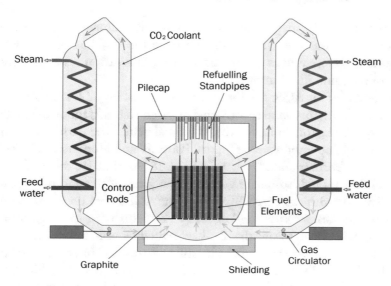

Fig. 22.3   MAGNOX Reactor

below). Wylfa was the last UK Magnox to shut down permanently, in 2015. The two exported MAGNOX stations (to Italy and Japan) had closed many years earlier.

The MAGNOX plants were low-temperature reactors—typically operating at around 360 °C. This gave them a similar thermal efficiency to PWRs. However, they were relatively large reactors and only able to achieve low fuel burn-ups. Many of the MAGNOX were able to be refuelled with the reactor running, which went some way to alleviating their high fuel throughput.

Other countries, including France and Spain, have operated a small number of reactors similar to MAGNOX, in the past. The only MAGNOX thought to still be operational in 2019 is the 5 MW Yongbyon reactor in North Korea, used primarily for weapons-grade plutonium production.

22.6   Advanced Gas-Cooled Reactors (AGRs)

In the UK, it was decided to build on the experience gained with graphite-moderated, gas-cooled reactors (the MAGNOX) but to re-design them for higher temperatures. This gave higher thermal efficiency and steam conditions similar to a coal-fired station, taking advantage of steam turbine developments. The 'Advanced Gas-cooled Reactors (AGRs)' are designed to run at around 600 °C, requiring a shift to uranium oxide fuel pellets, a bit like those in PWRs. Similarly, magnesium cladding was abandoned in favour of stainless steel. This was only made possible once the UK had gained access to enrichment technology as a stainless steel cladding captures too many neutrons for a natural uranium reactor.

A diagram of an AGR reactor is shown here as Fig. 22.4. Fourteen commercial AGR reactors were constructed at seven stations, following an experimental AGR reactor, built on the Windscale site. The earliest commercial AGRs to begin operation were Hinkley Point 'B' and Hunterston 'B' in 1976. The last were Torness and Heysham 2, in 1988. As of 2019, the AGRs are still all in service, though some are within a few years of the end of their operational lives. As with the MAGNOX fleet, the AGRs were designed to be refuelled with the reactor running 'on-load'. Problems with fuelling equipment and fuel construction meant that on-load refuelling was abandoned at a number of the stations, with the remainder still able to refuel but at reduced power.

Both MAGNOX and AGRs suffer from an inherent instability that is absent in PWRs and BWRs. That is, they have a positive moderator temperature coefficient. In both designs, the graphite expands little with operational temperature changes in the reactor, so unlike water, there is no reduction of moderation as the reactor heats-up. Most of the AGR reactors were, in any case, actually designed with more moderator than they need ('over-moderated') to compensate for the gradual loss of graphite from chemical processes over the life of the core. So what causes the positive moderator temperature coefficient?

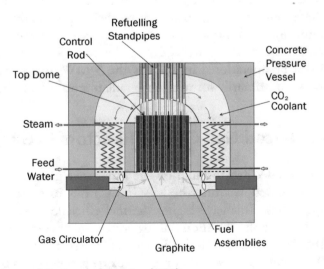

Fig. 22.4 Advanced Gas-cooled Reactor (AGR)

You'll probably remember the 'Resonance Capture Peaks' in uranium-238 (if not, look back at Chap. 9). Well, it turns out that plutonium-239 has 'Resonance Fission Peaks' at energies that are a little above typical thermal neutron energies. In other words, if the graphite moderator in a MAGNOX or AGR heats up a little (giving higher energy neutrons), with Pu-239 present in the irradiated fuel, the plutonium fission rate (power) will rise. The more power you have, the more the graphite temperature will rise, so you'll get more Pu-239 fission, etc.... It's a positive feedback effect or an inherent instability in the reactor physics. Incidentally, you don't see this effect in a PWR because it is under-moderated; the neutrons are already at higher energies, so this effect is swamped by the effects of the density changes in the water.

How is it possible to run a reactor with positive feedback between temperature and power? The answer is a combination of mass and time. There are hundreds of tonnes of graphite in a MAGNOX or AGR reactor, with only a limited surface exposed to the hot CO_2. It takes time (typically minutes) for it to heat up, so there's plenty of time for the operator, or an automatic control system, to change reactivity by moving control rods. To give you an idea: even though the reactors were inherently unstable, a number of the MAGNOX plants operated for their entire lives with their control rods under manual control. The operator would nudge them in or out every few minutes in response to changes (up or down) in reactor power. It was that easy!

The AGRs were designed with a feature to reduce the positive effects of moderator temperature. This is the re-entrant flow that you can see in Fig. 22.4 (the small blue arrows), with colder CO_2 being directed down between the fuel and the bulk of the graphite to keep the graphite temperature more uniform as power changes. Despite this, and unlike a PWR, active control of the AGRs by control rod movement is required, almost continuously.

In their early years, the AGRs performed poorly with low reliability and several operational problems, especially in fuel handling. Consequently, in the 1980s, the UK decided to switch to the more common PWR technology. Since then, AGR performance has significantly improved.

22.7 RBMK Reactors

The Chernobyl nuclear power plant comprised four RBMK reactors—'Reaktor Bolshoy Moshchnosti Kanalnyy', which translates as 'High Power Channel-type Reactor'. A bit like a MAGNOX or an AGR, the RBMKs are graphite-moderated reactors with low enriched fuel (~2% U-235). However, the fuel in an RBMK is located in pressure tubes within the graphite. Unlike

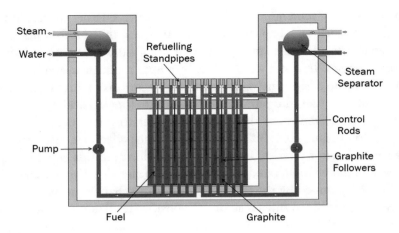

Fig. 22.5 RBMK reactor

in a MAGNOX or AGR, the coolant used is light water. The coolant water is allowed to boil within the channels (as in a BWR) before being separated and sent to drive the associated steam turbines. A little bit of everything.... (Fig. 22.5)

RBMKs clearly have a number of advantages over other reactor designs. They use light water as a coolant and only low enriched uranium fuel. Pressure tubes are relatively easy to make and can be replaced, so no large pressure vessel forging is required. Graphite is an inexpensive moderator, and it's proven possible to scale-up the RBMK reactors by adding additional channels. The largest RBMKs each produced 4800 MW of heat, converted into approximately 1500 MW of electricity; for a while, they were the most powerful reactors in the world.

On a much smaller scale, the first Russian nuclear electricity was produced by the 5 MW Obninsk reactor, a primitive RBMK, in 1954. This was two years ahead of the Calder Hall reactors but was kept secret by the Soviet Union, so Britain believed for a long time that it had been the first!

Unfortunately, this combination of water coolant and graphite moderator is inherently unstable. I've been told by visitors to working RBMKs that the operators in the control room are kept very busy, continuously adjusting control rods and pumps, simply keeping the reactor within limits. This instability eventually led to the disaster at Chernobyl (as described in Chap. 9).

Seventeen RBMKs were built by the Soviet Union including two in Lithuania and four in Ukraine. Of these, ten are still operational in 2019, all of them in Russia. They have each undergone many safety improvements, including the poisoning of the fuel with erbium (to reduce the positive void coefficient) together with improvements to the design of the control rod systems.

Outside of the Soviet Union, the only similar reactors of which I'm aware were the original Hanford Piles in America. These were natural uranium fuelled, water-cooled, graphite-moderated reactors but ran at low temperatures as they were used for plutonium production rather than for electricity generation.

From this point on, I won't include diagrams as there are just too many different designs to choose from.

22.8   Fast Reactors

Using a moderator isn't the only way to get a reactor running. If you put enough enriched uranium or plutonium close enough together you can achieve criticality on fissions caused by fast neutrons alone. This is the basis of all 'Fast Reactors'. Note that 'Fast' in this case doesn't imply instability. Fast reactors are dependent on delayed neutrons to achieve criticality in the same way as thermal neutron reactors. They have strong negative fuel temperature coefficients, and can also be designed such that the expansion of the fuel with temperature will cause enough of a change in geometry to give additional negative reactivity feedback.

Typically fast reactors use fuel at ~20% enrichment. Their cores are small and can have much higher power density than a PWR. Most fast reactors are built using a liquid metal coolant. Some use lead, but more usually it's sodium or a sodium/potassium mixture. Liquid metals have very high heat conductivity, and all three of these options have a sufficiently low melting temperature to be useful in a reactor. None of them captures significant numbers of neutrons though sodium will capture some, giving highly radioactive sodium-24 in the coolant. The really good thing about liquid metal coolants is that they don't require a primary circuit to be built to withstand high pressure; the reactor can run at almost atmospheric conditions, usually with a low pressure inert gas covering to the liquid metal.

The complexity of fast reactors rises once you get to the steam generators. Liquid metal coolants such as sodium and potassium (but not lead) react explosively with water, so it'd be a significant risk to separate them with single-walled steam generator tubes. It's more common to find double-walled tubes with an inert gas (e.g. argon) filling the gap. The engineering is more complicated, the heat transfer isn't as good, and there can still be leaks/problems.

At this point, you might be asking yourself 'Why bother?' The answer for many years was 'Fuel Supply'. Fast reactors have a very high neutron flux. If you put a blanket of uranium-238 around the fast reactor core, you will find that you can manufacture (or 'breed') more plutonium than the core is using. In theory, our existing uranium stocks could be stretched out to supply electricity for tens of thousands of years, if they were used in fast breeder reactors. This is why fast reactors were pursued enthusiastically by a few countries for many years.

Two Liquid metal cooled fast reactors were built at Dounreay in Scotland; both are now shutdown. In France three fast reactors were built, all now shut down. In the USA, eight fast reactors were previously operated, including a 400 MW (heat output) reactor at Hanford. In Japan, a single fast reactor (Monju) had a much-checkered career before final closure.

It's not entirely a negative picture though—both Russia and India currently have active liquid metal fast reactor programs, including both Submarine reactors (Russian) and electricity-producing reactors (both countries). The Chinese also appear to be keeping the fast reactor option open, with the construction of a large (600 MW electrical) sodium-cooled reactor at Xiapu.

22.9   Thorium

In India, the drive for fast reactors is tied-up with their plans to use their abundant reserves of thorium as reactor fuel. Natural thorium-232 can be changed into fissionable Uranium-233 by using thorium as a blanketing material in the same way as the uranium-238 in the fast breeder reactors, described above. That is, Th-232 can capture a neutron to become Th-233. Th-233 then decays to U-233.

Thorium gets a lot of positive press for being an alternative reactor fuel; usually, from people who don't understand that you have to run it through (or at least around) an already running reactor before it's of any use. Then you need to reprocess the fuel to separate the thorium from the highly radioactive fission products that will have arisen in the blanket (some of the uranium-233 will have fissioned). For that, you are going to need a large fuel reprocessing plant.

One of the other claims made by thorium enthusiasts is the inability to make weapons from thorium. Admittedly weapons development is not as easy as it is from a uranium-fuelled reactor as thorium doesn't produce any plutonium, and the U-233 is often contaminated with U-232… but it can be done. America, Russia and India are believed to have each exploded devices containing, at least in part, U-233 material derived from thorium.

22.10   The Paper Reactors

There are many more reactor designs sitting on people desks and computers than will ever be built. Some of these require sophisticated engineering with materials that are not yet developed. Or perhaps the issue is simply one of money—you only build what your company or country can afford when it can afford it; everything that gets designed in the meantime gets forgotten.

That's not to say that these alternatives don't have enthusiasts! My only advice on this would be to research what you hear very carefully. Sometimes the enthusiasts aren't nuclear scientists and will make claims for their reactor designs that can't possibly be true. At present, there are two groups of alternative-reactor enthusiasts actively lobbying for your attention: the Small Modular Reactor and the Molten Salt Reactor enthusiasts.

Small Modular Reactors (SMRs) are an attempt to overcome one of the biggest challenges in modern nuclear power station design: the cost. A full-scale (say, 1600 MW electrical) nuclear power station might cost you as much as £10 billion pounds, and you'll get nothing back on your investment until construction has finished and the plant is online. What if you could make smaller (modular) reactors? Would this work out cheaper and faster? It might. Or you might find that all of the other costs associated with running a nuclear power station—licencing, maintenance, and emergency schemes etc.—mean that there isn't that much difference on a per kWhour basis. A few countries are giving these SMRs a chance: notably China and Argentina. Canada, the UK and the USA may follow, but there's no guarantee.

Molten salt reactor designs are based on (mostly successful) experiments from the 1950s and 1960s. They come in two types. The less radical design uses solid fuel with a molten 'salt' as a coolant. 'Salt' in this context is used in the chemical sense, i.e. an ionic compound, such as lithium fluoride, beryllium fluoride etc.. Molten salt reactor coolants are often planned as a mixture of salts to reduce the melting temperature. In theory such a reactor could operate at a higher temperature than a PWR, so could be more efficient; though the chemistry is going to be tough.

The alternative, more radical molten salt reactor designs use a molten fuel circulating around the primary coolant. In such a design, criticality is avoided (except where you want it) by geometry and/or the lack of a moderating material anywhere except in your reactor vessel.

Molten salt reactor enthusiasts—especially for the second kind—will tell you that the reactor can't 'melt-down' because it is already liquid…. This is pure bunkum. Decay heat doesn't go away, simply because you're using liquid fuel. If you fail to remove the decay heat, the molten fuel will keep getting hotter and hotter until it burns through or melts whatever container you have it in!

More significantly a molten fuel salt reactor will require you to pump all of your highly radioactive fission products around the primary circuit. On a PWR, a single leaking fuel pin (out of 50,000 pins) can measurably increase doses to operators at the next refuelling shutdown (see the next chapter). What would it be like if *all* of the fission products were dissolved in the coolant?!

The enthusiasts will suggest that the coolant can be cleaned-up, removing the fission products, but there are dozens of elements involved, so the number of different chemical compounds that could be present in the molten salt is huge. I've not seen any clean-up technology that could do this, and the hot liquid coolant will be so radioactive that the clean-up plant itself would be horrendous. Personally, I'd be surprised if a successful full-scale molten fuel salt reactor could ever be practically operated or maintained, but the Chinese are attempting to build both solid-fuelled and molten-fuelled salt reactors—so I could be proved wrong!

22.11   And the Winner Is?

So what are countries actually building now, in 2019? Well, there are around 50 nuclear power plants under construction. With just a couple of exceptions, they are all large PWRs, BWRs and PHWRs. But please bear in mind that this is a 'snapshot'. The situation could look quite different in a few years.

22.12   Don't Just Take My Word for It…

OK, at this point I'll suggest that I'm likely to be biased. I started with MAGNOX reactors, side-stepped AGRs and moved to a large-scale PWR, where I've spent the last 25 years. I can understand why countries are choosing to build large PWRs, BWRs, CANDUs (or similar). I'm interested to see how lead-cooled fast reactors and thorium fuel cycles develop. At the moment, I can't see the need for either SMRs or Molten salt reactors. But that's just my point-of-view.

Everyone in the nuclear industry has their own preferences. Look at the evidence, research different points of view, challenge claims that look to good to be true, think about the practicalities (e.g. radiation dose) from maintenance and refuelling. Then, make up your own mind.

23

How to Build Your Own Reactor

Don't!

No, really. Don't even think about it.

If you managed it, you would be risking your health and that of your neighbours.

You get the idea…? Sure?

OK, but what if you *were* that way inclined, how might you go about building a small nuclear reactor?

23.1 First the Fuel

You are only going to be able to build a working reactor if you can find some material that will readily undergo fission. Practically speaking, that's going to be either uranium or plutonium. There are some more exotic fissionable materials out there, but they tend to exist in laboratories in quantities too small to fuel a reactor. You might have been thinking 'Thorium', but as you've already seen, that'll only be of any use if you first have a running reactor to put it in.

23.2 Plutonium

Where might you find some plutonium?

There's a surprisingly large amount of plutonium in the world. This is because we've been running lots of uranium fuelled reactors for many decades,

© Springer Nature Switzerland AG 2019
C. Tucker, *How to Drive a Nuclear Reactor*, Springer Praxis Books,
https://doi.org/10.1007/978-3-030-33876-3_23

and, as you've seen, uranium-238 can be turned into plutonium-239 inside a reactor.

Apart from a few small samples used in research, most plutonium is going to found in one of three places:

- Still held inside irradiated fuel, within or outside of a reactor, where the fission products around it are going to mean that it is lethally radioactive.
- Stored after reprocessing fuel. Sellafield, for example, holds a stock of around 140 tonnes of plutonium from historic reprocessing activities. Countries (not individuals) run reprocessing plants, and I'm guessing these sites are going to be pretty secure.
- Held by the military either as nuclear weapons or as fuel for submarines or ships. The key word there is 'military'; again, very secure.

Honestly, getting hold of enough plutonium to build a reactor (a few tonnes) is not something you're likely to accomplish as an individual.

23.3 Enriched Uranium

Ideally, you'd be looking to find a source of enriched uranium as that would make a smaller reactor—you'd need less fuel. The more enriched the uranium, the more significant the benefit in scale. But, Highly Enriched Uranium (HEU), at 80–90% U-235, is effectively weapons material, so is guarded in the same way as plutonium; you're not going to find any of that. Medium enrichments are quite rare, used only in specialist reactors, so cross those off your list as well.

On the other hand, Low Enriched Uranium (LEU), say up to 5% U-235, is common as many reactor designs, including PWRs, BWRs, and AGRs all use LEU as fuel. That also means that manufacturers of LEU fuel are relatively common worldwide. So could you just go out and buy some LEU?

Probably not.

23.4 The Nuclear Non-Proliferation Treaty (NNPT)

It's time for a bit more history. As you know, the first nuclear reactors were developed to produce plutonium for weapons. To begin with the technology for these reactors, the weapons, and for uranium enrichment plants, was controlled by the USA. Due to espionage and development work by Russian scientists, this had clearly changed by the time the USSR detonated its first nuclear weapon in 1949. Over the next couple of decades there was an increasing concern that an all-out nuclear war could be triggered between the USA and the USSR. In the meantime the UK detonated its first nuclear device in 1952, France in 1960 and China in 1964.

Skipping over a lot of politics and negotiation, an international "Treaty on the Non-Proliferation of Nuclear Weapons" was opened for signature in 1968. The treaty has three aims:

1. Non-proliferation: Countries who have nuclear weapons pledge not to share nuclear weapons technology with those who don't. Countries without nuclear weapons pledge not to try to develop or acquire them, and also to accept 'Safeguards' inspections from the International Atomic Energy Agency (IAEA) to verify that they are adhering to the treaty. Signatory states with nuclear weapons also either submit voluntarily to IAEA inspections or have demonstrated an equivalent safeguards system to the IAEA's satisfaction.
2. Disarmament: Countries with nuclear weapons undertake to negotiate to reduce or remove their nuclear arsenals. (I admit that this doesn't look to have been very successful, though some reductions have been made.)
3. Peaceful use of Nuclear Technology: All countries have the right to develop nuclear energy programs, subject to their obligations under the NPPT, and to co-operate with other signatories to the treaty in such development.

Four states have never signed the NPPT: India, Israel, Pakistan and South Sudan. North Korea signed then withdrew. However, most countries have signed and adhere strictly to the requirements of the NPPT. Any country found to be breaching the treaty risks being left out in the cold concerning the supply of nuclear technology and fuel.

What this means in practice, is that in most countries, companies and state-run organisations that produce LEU will legally only be able to sell it to other companies and organisations that have inspection arrangements under the NPPT.

Of course, if you're a large company developing or building nuclear reactors, then you'll already be spending millions of pounds to satisfy your country's nuclear regulators that you have a safe design. Entering into an acceptable safeguards inspection regime will probably be quite straightforward for you. However, as an individual wanting to build your own nuclear reactor, you're not going to be able to do this, so you won't get a look-in. No enrichment company will risk its own inspection status by selling to someone outside the treaty.

23.5 Natural Uranium

This is the final option: find somewhere where uranium exists naturally in the ground and dig it up.

Uranium is surprisingly common—it's about 40 times as abundant as silver in the Earth's crust. Typically though, it occurs at very low concentrations, just a few parts per million, making extraction impractical. Of course, as with many other minerals, there are places where it is more concentrated, and minerals such as pitchblende can be mined, containing a few per cent uranium oxide.

Unsurprisingly, countries with large reserves of minerals containing uranium are the same countries that mine and export uranium to everyone else. The largest exporters are Kazakhstan, Canada and Australia with more than 50,000 tonnes of uranium being mined worldwide each year from these, and around a dozen other countries.

Again, each of these countries and mining companies will (in theory) only sell uranium to signatories of the NPPT. However, let's assume that you are a lucky land-owner and have direct access to an ore deposit, together with the ability to smelt the ore into reasonably pure metal. How much would you need? Think back to the description of Chicago Pile 1 (CP-1) in Chap. 4... that reactor used around 50 tonnes of uranium metal. If you had a good ore at 1% uranium content, that would mean digging up and processing 5000 tonnes of rock. How big is your wheelbarrow?

Even if you manage to accumulate this much natural uranium, what are you going to use as a moderator? You can't use normal (light) water as a moderator with natural uranium fuel as light water captures too many neutrons—reactors moderated with light water will only work with enriched uranium or plutonium fuel. In theory, you could use heavy water as an alternative moderator with natural uranium, but heavy water is expensive—roughly ten times the price of a decent single malt whiskey—and you'd need tonnes of it. Oh, and heavy water is subject to the same NPPT inspection regime as uranium and plutonium, so you are going to struggle to find a supplier.

I guess that means that you're probably stuck with using graphite as a moderator. The key problem with natural uranium fuelled, graphite-moderated reactors, is that they need to be big. If you build a small one, the leakage of neutrons from the sides will be too high and you'll never achieve criticality. As an example, CP-1 was made from 360 tonnes of graphite blocks. Where are you going to get this much high-purity graphite?

23.6   It's Not Going to Happen

You can probably see by now that, if you're an individual, there is no practical way of building your own reactor. Even if you're an individual with lots of money and manpower, you're likely to run up against restrictions in the supply of fuel, moderator or other technologies. You simply won't be able to buy what you need.

It's worth noting that in most countries, building your own reactor would, in any case, be illegal. In the UK, for example, legislation requires a 'Nuclear Site Licence' to be granted for any fixed nuclear reactor (cleverly excluding nuclear submarines!). A nuclear site licence cannot (by law) be issued to an individual—only to a Company—and would only be issued if the full set of 'Site Licence Conditions' (see Chap. 17) could be demonstrated to be met. You're not in that league.

23.7   Has Anyone Ever Tried?

Surprisingly, yes.

David Hahn, an American Boy Scout, managed to persuade a smoke detector company that he needed 100 detectors for a school project. Having bought these at a discount, he dismantled them to extract the radioactive source in each one. Most smoke detectors contain a tiny quantity (less than a microgram) of americium-241. This a long-lived radioactive material which decays by emitting alpha particles and gamma rays. Inside a smoke detector the alpha particles travel a short distance through the air before reaching a detector. If the air contains smoke, fewer alpha particles will reach the detector, and this drop in signal will cause the smoke alarm to sound. Outside of a smoke detector, the americium simply acts as a source of alpha particles.

David Hahn increased his inventory of alpha-producing material by scraping radium from old luminous dial watches and clocks he'd bought in antique shops. Alpha particles on their own won't drive a nuclear reactor, but if you include some beryllium in your design (a friend stole him a strip of beryllium from the local community college's chemistry laboratory), you can produce neutrons—beryllium-9 will capture an incoming alpha particle (provided by the americium and radium) to become carbon-12, emitting a neutron in the process. Incidentally, beryllium is extremely toxic!

The final step was to encase his neutron source in a blanket made up of uranium (he'd ordered a small amount from Czechoslovakia, posing as a college professor) and thorium which he'd extracted from hundreds of camping lamp gas mantles. He had aimed to use the neutrons to convert the thorium-232 into the fissile isotope uranium-233, and then… well, who knows? As a laboratory experiment it sounds fascinating; but in a garden shed, in the real world, he'd constructed a horribly radioactive device, contaminating his house and surrounding area with radioactive materials. The clean-up cost tens of thousands of dollars.

His idea has been copied by a few others around the world, but it's never ended well.

So, don't try it.

No, really. Don't even think about it.

24

And There's More...

24.1   One Small Book

Nuclear energy is a big subject. It would be foolish to try to cover everything in one book, so instead I've concentrated on the safe operation of a PWR power station. This subject alone is easily sufficient to have filled most of the previous chapters.

However, before I finish, I think it's sensible to spend just a few pages on some other aspects of nuclear energy. Then, if you want to, you'll have some pointers on areas that you might wish to research further.

24.2   Not Just Operations

I'll be honest, control room operators make up less than 10% of the staff of your power station, even though they work in shift teams to provide 24-hour cover. They are assisted by a similar number of plant operators—the people who actually go onto plant and operate valves, test pumps etc. The plant operators' training is different from that of the control room operators, but they are an essential part of operations. Not everything can be done from the control room.

Your station will have technical staff, able to advise the control room (and others). These will include specialists in radiological protection who advise and place limits on work that involves potential doses of radiation. Similarly, industrial safety and fire safety specialists will advise on possible risks from the sort of engineering work you'd expect on a sizeable industrial plant.

© Springer Nature Switzerland AG 2019
C. Tucker, *How to Drive a Nuclear Reactor*, Springer Praxis Books,
https://doi.org/10.1007/978-3-030-33876-3_24

I've mentioned chemistry several times in this book, so you'll not be surprised that there are a lot of chemists sampling, dosing and advising, to optimise the lifetime of the plant. In terms of technical advice, your station will also employ a number of nuclear safety or safety case specialists. As you've seen the safety case for a modern nuclear power station is going to be enormous. You are going to need specialists in this area to ensure that the plant is always operated within the assumptions, limits and conditions of the safety case, and at an optimally low risk to the public and staff. These nuclear safety specialists are the ones that are going to write the 'rule book' for your control room staff to follow.

It doesn't stop there. There are more than 200,000 plant items in your power station. Think of how many engineers and maintenance staff it's going to take to keep all that equipment in good order: tested, lubricated, maintained and, when life-expired, replaced. Engineers come in different flavours—you'll need mechanical engineers, civil engineers, instrumentation engineers, electrical engineers etc, and beyond the engineers on-site, perhaps some other specialists serving the wider fleet of power stations from an engineering headquarters.

Not everyone is a scientist, operator or engineer. You've seen in an earlier chapter how a refuelling outage might include 10,000 separate tasks. Those all need to be planned and put into procedures long before the outage starts. Outages employ a large number of contractors (perhaps 1000) and use up a lot of spares. You are going to need supply chain specialists to deal with all of that. On a smaller but more relentless scale, maintenance and operations staff together might have to get through 500 or so tasks every week while the reactor is running. Again, this all needs to be planned so that nothing is forgotten. Nuclear power stations have a lot of planners!

Finally, there are going to be the kinds of roles that you find in any large organisation: human resources, finance, catering, cleaning, security etc. But I think there's a difference between these roles at a nuclear station and elsewhere. A nuclear human resources manager might well be found putting on a set of overalls and visiting the plant. A fresh pair of eyes can work wonders in spotting problems that operators who visit every day might not have noticed, and everyone is encouraged to do this. Similarly, a cleaner is far more likely to see (and report) a dripping pipe than an engineer—because they'll be the one looking at the floor to clean it. At a nuclear power station, everyone has a stake in looking after the plant and keeping it safe, regardless of their job title.

Why go through all of these roles? To show you that there are very many possible careers in the industry; you don't have to be a physicist or engineer to contribute.

24.3   Spent Fuel...

This books' description of irradiated fuel stops as the fuel reaches your fuel storage pond. That may have prompted the question 'Then what?' The answer depends on where you live and what kind of reactor you're running. In the UK, current government policy is to store irradiated PWR fuel above ground until a deep underground repository is up and running in a few decades. The spent fuel would then be packaged into canisters and stored indefinitely underground. At Sizewell 'B', spent fuel is presently being moved out of the fuel storage pond after a few years decay, into dry, shielded casks for medium-term storage, awaiting the underground repository.

In contrast, up until recently, fuel from the UK's Advanced Gas-cooled Reactors and MAGNOX reactors was reprocessed at Sellafield. Reprocessing is just jargon for chopping up the fuel and dissolving it in acids. This enables you to separate out the unused uranium, the plutonium and the radioactive fission products which can then be processed into an immobile glass (vitrified). Reprocessing is a great way of reducing the volume of radioactive waste, but the downsides include the cost, the waste streams that are created by the processing and the liberation of the plutonium that could then be diverted into weapons.

It's this last aspect (plutonium & weapons proliferation) that makes reprocessing so contentious internationally, and shows up sharp divides in policy between countries. France sends fuel from all 58 of its PWRs to its La Hague site for reprocessing. In contrast, the USA reprocesses none of the fuel from its near 100 commercial PWRs and BWRs. After decades of reprocessing, the UK has decided to stop reprocessing AGR fuel—moving instead to storage then underground disposal. MAGNOX fuel has to be reprocessed as it can't be stored indefinitely once irradiated (it dissolves slowly in fuel storage ponds). All of the UK MAGNOX reactors have now shut down, so this reprocessing activity will be time-limited. Other countries have made their own choices.

24.4   ...and Radioactive Wastes

Putting fuel to one side, there are other less hazardous waste streams that have to dealt with on a running PWR. Some solid wastes such as filters, with short-lived, low-level radioactive contamination, can be disposed of at shallow underground sites (such as Drigg in Cumbria). Longer lived, or more active wastes will have to be stored above ground (either at places such as Sellafield or at the power stations themselves) until an underground repository is available.

The situation for radioactive liquids and gases is more complicated. These can be cleaned up to a great extent, but removing some radioactive chemicals, such as tritium (radioactive hydrogen-3) is impractical. It's therefore necessary to release small amounts of radioactive material into the environment. This is heavily regulated and carefully measured. The natural world is already radio-active—through both natural processes and man-made contributions. The idea of regulated discharges is to ensure that there is no measurable increase in environmental radiation levels, relying on the liquids and gases being very quickly dispersed in the atmosphere or, usually for liquids, the cooling water (sea). Discharges at even these low levels have their critics, but there is an overwhelming body of evidence to demonstrate that this presents no appre-ciable risk to the environment or to the public.

24.5   At the End of the Day

All power stations wear out. Experience worldwide shows that for nuclear power stations, this is often decades beyond the lifetime the original designers imagined. I think this simply reflects the ongoing improvements in mainte-nance, inspection and replacement capability. If you have a nuclear power station with a worn-out component, it's usually much more rewarding finan-cially to replace that component (even at the cost of millions of pounds) and then continue generating, rather than to shut down the plant permanently.

Even so, plants do eventually become uneconomic, are shut down and enter the stage of life we call 'decommissioning'. Again jargon: all it means is 'clean it up and knock it down'. In the UK we've found that our MAGNOX reactors (and to a lesser extent the AGRs) were not designed with decommis-sioning in mind. They were large and were constructed of materials that, once irradiated, take a long time to decay away to safe levels. It's not uncommon to hear people talk about MAGNOX plants taking 100 years and billions of pounds to decommission.

In contrast, the experience of PWR decommissioning is that it can be per-formed much more quickly—achieving a green-field site in around 10–15 years. It's also a good deal cheaper and can be easily funded by setting aside a small proportion of revenue from electricity generation, into a pro-tected fund. In the UK this fund is known as the Nuclear Liabilities Fund (NLF), and it is administered by government. The radioactive parts of a PWR power station are (relatively) very small compared with MAGNOX and AGR

designs. In addition, the design/construction of a PWR lends itself to dismantling. It's hardly surprising that decommissioning is going to be easier.
Remember that a modern PWR might run for 80 years (or more), producing
more than a million pounds worth of electricity every day that it operates.
Decommissioning costs are going to be a tiny fraction of that revenue.

24.6 Off the Grid?

Where electricity production in the UK used to be dominated by large coal-
fired power stations, these have now mostly been shut down or operate at only
peak demand periods. Instead there's now a preponderance of smaller gas-
fired stations, mixed in with nuclear and renewables such as wind and solar.
Nuclear stations can now find themselves to be the largest individual generators on the grid at periods of low demand. Put another way, an unplanned
shutdown or trip of your reactor can have a significant impact on a grid that
has lost most of its other large generators.

The UK isn't the only country struggling with this change in generation—
away from coal, towards low-carbon (semi-unpredictable) renewables. Most
countries in Europe are part of an interconnected grid system, so easing the
shock from a single reactor trip. The UK is currently building more intercon-
nectors, but with larger stations possibly coming on stream in a few years, I
suspect that this threat to the grid will persist for a few more years yet. It's a
difficult time to be running an electricity grid; one of the unintended consequences of phasing-out coal.

24.7 Books, Accidents and Weapons

The first comprehensive book I read on nuclear energy was 'Nuclear Power' by
Walter C. Patterson. I found it pretty close to impartial despite being written
by, as I discovered later, an anti-nuclear campaigner. It's got a North American
bias and is now out of print, but is still available on-line or second-hand. I'd
still recommend it.

If you're interested in the development of the first nuclear reactors, then
you'll find yourself reading about the development of the first nuclear weapons as the two are inextricably linked. A detailed description can be found in
'The Making of the Atomic Bomb' (by Richard Rhodes). It's a long read but
well worth it.

There are some excellent books out there on nuclear accidents: I'd recommend 'On the Brink: The Inside Story of Fukushima Daiichi' (by Ryusho Kadota) for its first-hand accounts of the events following the Tsunami. If you work at a nuclear power station, this book often prompts the very personal question 'How would I react, in the same circumstances?' In contrast, but also well worth reading despite the occasional (minor) error in the physics, is 'Chernobyl: History of a Tragedy' (by Serhii Plokhy) which shows just how strongly the development of an accident can be influenced by the political regime under which it occurs.

You might be wondering why I'd be recommending books on nuclear accidents? It's the same idea as reading about railway accidents when you're a Signalman—if you don't learn about what can go wrong, how are you going to spot the warning signs and prevent a similar event? Your best chance of dealing with an accident is to avoid having one, and you'll only do that if you're fully aware of those that have already happened. In the world of nuclear energy, you sometimes hear it described as 'Operational Experience: Use it or Become it.'.

24.8   The Politics and the Campaigning

As I said at the start, I'm not going to attempt to describe or defend the politics of nuclear energy. There are hundreds of nuclear reactors already running in the world. Some of these have been operating for decades and are reaching the end of their useful lives (or already shutdown). Dozens more are currently being built. Regardless of your views on nuclear energy—a hazard to be avoided, or a useful source of low-carbon energy—it already exits, on a massive scale.

Many other authors have published books that are variously for or against nuclear energy. Sometimes this is obvious from the title; sometimes only from the reviews. Just occasionally, I find one that is pretty impartial. I'm not impartial. I'm an enthusiast for nuclear energy—something you've probably already guessed—so there'd be little point in me adding to the mix by only putting forward that point of view.

25

Conclusion

Let me remind you of the three key concepts I outlined, way back in Chap. 1.

- *Reactivity, or how the conditions inside the reactor affect the fission chain reaction.* I accept that there's a bit of physics to take on board before the idea of reactivity makes sense, but to a reactor operator, this is a vital concept. In this book you've seen the things that can affect the reactivity of your reactor, and how it will behave in response to reactivity changes. If you don't know the effect on reactivity from what you're about to do, then don't do it!
- *Reactor stability, the feedback mechanisms that hold it steady.* A bit more physics—for example, fuel and moderator temperature coefficients—and we have a stable reactor. Of course, stability is a double-edged sword. It holds the plant steady, which is great when you want it to be steady, but as you've also seen, you'll have to fight against it if you're going to move the plant around in power or temperature.
- *Plant stability, what happens when you connect your reactor to the rest of the plant (and beyond).* This is the real key. Most reactor physics textbooks stop at the reactor, but as you've seen in this book, the behaviour of a real PWR is dictated by what it's connected to. You move the valves on your turbine, and your reactor will follow it. You break a steam line, and your reactor power will go up to meet the higher perceived steam demand. You trip the turbine, and the reactor power will drop like a brick unless you have somewhere to dump the steam. And so on.

© Springer Nature Switzerland AG 2019
C. Tucker, *How to Drive a Nuclear Reactor*, Springer Praxis Books,
https://doi.org/10.1007/978-3-030-33876-3_25

There is, if you like, a fourth concept that's been present in the background of almost every chapter: 'Safety'. It's the sole subject of one of the more detailed chapters in this book (Chap. 17). Nuclear power stations provide enormous amounts of low-carbon electricity in a predictable, controllable way. They also present unique risks in terms of accidents leading to core damage and the potential release of radioactive fission products.

Anyone who works at a nuclear power station needs to start from the idea that 'Nuclear Safety is always our Overriding Priority'. They need to challenge any behaviour or condition that could take the plant away from a safe state or outside its intended operational envelope. It is possible—and indeed common—for nuclear power stations to be operated at very low levels of risk. It has to be that way, but it doesn't happen by chance.

As you've seen, the role of a reactor operator on a PWR, for most of the time, isn't to drive the reactor itself—though you'll have to do that sometimes, when taking the reactor critical for example. Instead, their role is usually to control the plant as a whole, to ensure that the reactor stays within the assumptions of the safety case. An operator also needs to be ready to respond to any eventuality whether it's from inside or outside the plant. That's why the final figure in this book isn't a glossy picture of a reactor; it's a diagram of the primary and secondary circuits *combined* (Fig. 25.1). That's the mental model you need to take away with you to be a successful PWR reactor operator.

And finally:

If you've read this book all the way through, you'll have seen that PWRs are inherently simple machines. But, you'll have also encountered some pretty tricky concepts when it comes to PWR physics, behaviour and control. If all of this made sense to you at first reading, then well done! If not didn't, don't worry; personally, it took me years to get the hang of xenon behaviour and of steam generator stratification. I'm still not sure about MVARs...

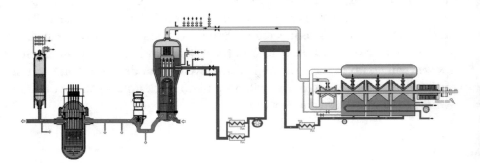

Fig. 25.1 Primary and secondary circuits

What I hope you've gained from this book is an appreciation of what it really means 'to drive a nuclear reactor'. PWRs *are* stable, but they need to be overseen by highly trained, responsible operators, backed-up by specialists in many disciplines.

That's what keeps them Safe.

Picture Credits and Sources

The photographs in this book of reactor fuel, plant items, control room panels and controls have all been reproduced with the kind permission of EDF Energy, including approval from a security perspective.

Similarly, the cutaway diagrams of the Primary Circuit (3D), Reactor Pressure Vessel, Pressuriser, Steam Generators and Moisture Separator Reheater are taken from EDF Energy training material. The various diagrams of the Primary and Secondary Circuits (including turbines) have been adapted—albeit heavily—from similar EDF Energy training material.

The drawing of Chicago Pile 1 (CP-1), together with the flux trace taken from its first criticality, are included in this book under a "free use with attribution" license from the Argonne National Laboratory. The photograph of my CP-1 'souvenir' was taken by Lynette Tucker.

The only quoted text within the book is that of physicist Herbert Anderson and his description of the initial criticality of CP-1. This appears in numerous sources, including Richard Rhodes book "The Making of the Atomic Bomb', mentioned in Chap. 24.

UK Nuclear Site Licence conditions are available on the UK Office for Nuclear Regulation (ONR) website. Similarly, the full text of the Nuclear Non-Proliferation Treaty is available from the International Atomic Energy Agency (I have summarised its key points within the book).

The boiling point (saturation) curve and fission product distribution graph are taken from generic data available on-line. UK electricity demand figures

© Springer Nature Switzerland AG 2019
C. Tucker, *How to Drive a Nuclear Reactor*, Springer Praxis Books,
https://doi.org/10.1007/978-3-030-33876-3

are downloadable from the 'Gridwatch' website, amongst others. I have tried to use 'representative' data for the latter.

Transient graphs of, for example, xenon behaviour, together with representations of nuclei of different sizes (as per the one above), have been derived from software that has been specially written, by myself, for this purpose. Other graphs/diagrams, including sketches of alternative reactor designs, have been drafted free-hand using simple software.

All other text, graphs, flowcharts, diagrams and illustrations within this book are my own. They are drawn from my own experiences and discussions within the industry, together with some research and fact-checking on-line, especially when this concerned more distant events.

Index

© Springer Nature Switzerland AG 2019
C. Tucker, *How to Drive a Nuclear Reactor*, Springer Praxis Books,
https://doi.org/10.1007/978-3-030-33876-3

Printed in the United States
By Bookmasters